Methods in Molecular Biology

Series Editor
John M. Walker
School of Life and Medical Sciences
University of Hertfordshire
Hatfield, Hertfordshire, AL10 9AB, UK

For further volumes:
http://www.springer.com/series/7651

Molecular Genetics of Asthma

Edited by

María Isidoro-García

Department of Clinical Biochemistry, University Hospital of Salamanca, Salamanca, Spain; Department of Medicine, University of Salamanca, Salamanca, Spain; Salamanca Institute for Biomedical Research, IBSAL, Salamanca, Spain

Editor
María Isidoro-García
Department of Clinical Biochemistry
University Hospital of Salamanca
Salamanca, Spain

Department of Medicine
University of Salamanca
Salamanca, Spain

Salamanca Institute for Biomedical Research
IBSAL, Salamanca, Spain

ISSN 1064-3745 ISSN 1940-6029 (electronic)
Methods in Molecular Biology
ISBN 978-1-4939-3650-2 ISBN 978-1-4939-3652-6 (eBook)
DOI 10.1007/978-1-4939-3652-6

Library of Congress Control Number: 2016940979

Printed on acid-free paper

This Humana Press imprint is published by Springer Nature
The registered company is Springer Science+Business Media LLC New York

Preface

Asthma is a multifactorial disease characterized by bronchial hyperresponsiveness and inflammation associated with airway obstruction episodes. Genetics is considered to play an essential role in the etiopathogenesis of the disease. The aim of this book is to provide the scientific community with a useful resource to study the molecular genetics in asthma. This book will guide the reader from the basic application of molecular genetics to the more complex gene expression analysis by using different models of study. Since asthma is a complex disease, this book is designed in six different sections to provide a review of the most useful techniques with examples of their applications in specific laboratory protocols.

Section 1. Genomic Analysis

Asthma is a polygenic and phenotypically heterogeneous disease in which the environmental component and therefore the epigenetic contribution also appear to play an important role. Therefore, the genetic approach to the disease is complex, and to date a clear pattern of inheritance has not been described.

The scientific community has focused on the identification of candidate genes that mediate the disease. Different methodological approaches have been applied for understanding the mechanisms underlying asthma. These technologies are vertiginously changing in the last years. Although the development of new technologies has changed from single gene studies to whole genome approaches, the focus of such studies has always been the identification of biomarkers that inform about an increased disease risk.

The study of genetic variation has exponentially improved with the application of new technologies, providing a huge window of new information, in some cases beyond the human capacity for analysis and interpretation. In this sense, the parallel development of bioinformatics and biostatistical analysis provides reliable results about the vast amount of information obtained from "omics" studies.

The majority of the association studies of asthma have been focused on studying single nucleotide polymorphisms (SNPs), although nowadays the copy number variations (CNVs) are also being studied. The majority of SNPs have a high frequency in the human genome and a minimal impact on the disease. Nevertheless, depending on their location into the gene, some SNPs have functional consequences, such as changes in protein function or in the gene expression levels.

The results obtained in association studies of asthma have not always been replicated in different populations. In addition, in some cases, contradictory results have been obtained. There are several causes for this disparity, including deficient experimental quality measures, lack of rigor of statistical treatment, ethnic differences, or different clinical criteria for the selection of the study samples, among others. In addition, the phenotypic heterogeneity of asthma is an added problem.

One of the main conclusions of genomics studies is the need to separately analyze the phenotypes of asthma. It has been reported that some genetic variants are associated with more than one phenotype of asthma; however, most of the variation is associated with specific subgroups. In Chapter 1 of this book, the evolution of genetic and genomic analysis in asthma is described.

The rapid evolution of the molecular methodology requires specific validation techniques. To date, the capillary sequencing technique, based on the Sanger's method, continues to be the gold standard. This classic technique requires the usage of different bioinformatics platforms in order to validate genetic variations. Chapter 2 provides a review of "in silico" protocols used to characterize and validate new genetic variants.

Section 2. Gene Expression and Asthma

This section is devoted to the gene expression study. The section begins with Chapter 3, in which gene expression and the different techniques used for its study are described. It begins with Northern blot analysis and in situ hybridization, in which labeled nucleic acids are used to detect targets immobilized on a support. These techniques allow estimating the level of a messenger RNA (mRNA) and hence the gene expression level.

The next chapter is focused on real-time PCR (qPCR), established as the basic quantitative technique for the analysis of gene expression. This technique allows a much more accurate estimation of the amount of transcribed mRNA. Thus a detailed description of qPCR is given in Chapter 4. To perform this quantification, an initial reverse transcription over total mRNA is developed to obtain the corresponding cDNA. Subsequently, the amount of cDNA is quantified by a PCR with specific primers. The amount of emitted fluorescence is directly proportional to the amount of the cDNA template present in the sample and therefore proportional to the level of gene expression.

This technique has many advantages such as speed (the assay time is approximately one hour), simplicity (the assay requires few reagents), convenience (it does not require post-amplification processing), sensitivity (it discriminates very few copies of mRNA), specificity (a well-designed assay is specific for a single target gene), robustness (a well-designed trial will give results through a wide range of reaction conditions), high performance (thousands of reactions can be carried out in a single experiment), familiarity (the PCR is well known), and cost (the reagents are affordable).

On the other hand, real-time PCR has some disadvantages compared to more innovative techniques such as expression arrays or RNAseq. The main disadvantages would be that real-time PCR requires an independent reaction for each of the genes and a constitutive expression control (it should be known and with a stable gene expression). A review about the election of constitutive expression controls for qPCR assays is provided in Chapter 5.

The technological takeoff observed in recent decades has also enabled the implementation of new and more ambitious experimental approaches acting at the "omic" level, which allow the gene expression analysis of multiple genes in a single experiment. Among these techniques, the arrays were the first popularized. Arrays were initially applied to the study of DNA sequences but were soon adapted to the study of RNA.

The cDNA arrays were applied to the study of gene expression in many diseases, including asthma, with interesting results. However, as these studies were developed, the scientific community was aware of the need of carrying out experimental quality and statistical controls that have become essential to process and interpret the vast amounts of data obtained in these studies.

The set of information obtained in "omics" studies about the expression levels of all the genes analyzed in a sample and in a particular condition is called transcriptome. The transcriptional analysis has been widely employed to analyze biological functions, since most of the biochemical changes underlying the normal development or the disease states are determined by specific expression profiles. Transcriptome analysis has considerably improved with the development of new generation techniques such as expression microarrays and RNAseq.

RNAseq is based on a high-throughput sequencing technology, in which the total mRNA is firstly converted to cDNA and subsequently sequenced. Finally, the reads are mapped onto the reference genome sequence. The expression levels are quantified depending on the number of reads. In addition, by this method, splicing variants and transcription initiation sites can be identified.

Several advantages of the RNAseq vs. the microarrays have been described. RNAseq provides direct information of all polyadenylated mRNAs and so it is not limited by the probe design. Another advantage, with important biological implications, is that RNAseq characterizes much better isoforms and alternative splicing events. The RNAseq data also have multiple advantages compared to microarray data, such as more dynamic range, less noise, higher specificity, higher reproducibility, better yields of digital information, and the fact that they do not require normalization. In addition, RNAseq results have been described as extremely accurate to measure transcript levels as determined by qPCR validation.

Section 3. Study of Regulatory Mechanisms of Gene Expression Involved in Asthma

In previous chapters, gene association studies both at individual level and at whole genome level have been revised. Studies used to detect associations between changes in gene-level expression and disease development have also been reviewed in Section 2. In this section, molecular techniques used to detect the functional mechanisms by which these genetic changes lead to changes in gene expression or in function of genes that eventually trigger the disease will be reviewed.

In Chapter 7, the protocol for EMSA (electrophoretic mobility shift assay) studies is described. EMSA studies are employed to detect interactions between DNA sequences (usually fragments of the promoter) and proteins (usually transcription factors). The technique is based on the fact that the electrophoretic mobility of a nucleic acid becomes slower when it forms complexes with proteins than when it is free. Thus the detection of a change in the electrophoretic pattern caused by adding a specific protein, or a protein mixture, indicates the existence of one or several putative transcription factors that bind to the promoter sequence of interest.

The analyses in which these interactions are detected are usually confirmed by performing competition assays. The addition of an unlabeled promoter fragment should remove the detected binding complex, indicating that the binding is specific.

Once the presence of DNA–protein complexes is confirmed, they can be isolated from the corresponding filters, and the bands of interest can be used to isolate the specific protein responsible for binding, i.e., the potential transcription factor that binds and regulates this promoter sequence.

These assays have also been used to analyze in parallel the binding capacity of different versions of a promoter region with different genetic variations. The variants in which a change in the electrophoretic mobility is detected would be responsible for changes in the transcription factors binding and thus could be responsible for changes in the corresponding gene expression. Generally, the results obtained in these studies are correlated with gene expression studies.

Other ways to analyze the mechanisms by which the binding of regulatory proteins and DNA leads to functional changes are Western blot studies (Chapter 8) and studies of chromatin immunoprecipitation (Chapter 9). In Western blot studies, proteins are separated from a protein mixture by acrylamide gel electrophoresis. Subsequently, proteins are transferred to a membrane where a specific protein can be identified by using a specific antibody.

Chromatin immunoprecipitation (ChIP) is a powerful technique that allows analyzing protein–DNA interactions in vivo. From a chromatin preparation, a protein of interest is selectively immunoprecipitated to determine the DNA sequences associated. ChIP can be used to determine whether a transcription factor interacts with a candidate target gene and to map the localization of histones with post-translational modifications on the genome. Several modifications have been developed such as ChIP-chip and ChIP-seq. The ChIP-chip method combines chromatin immunoprecipitation and DNA microarray analysis to identify protein–DNA interactions in vivo. ChIP-seq consists in the immunoprecipitation of chromatin followed by high-throughput sequencing.

Another method to study gene expression regulation is through RNA interference (RNAi). A detailed guide for gene silencing mediated by siRNA is described in Chapter 10. It is focused on the most used methods: lipid-mediated and electroporation transfections. The RNAi technique is based on the mechanism used by many organisms to silence gene expression through RNA molecules of a targeted gene with high specificity and selectivity. Different types of small ribonucleic acid molecules, such as microRNA (miRNA), small interfering RNA (siRNA), and short hairpin RNA (shRNA), are involved in RNA interference. RNAi is a relevant research tool in cell cultures and in vivo experiments, because the synthetic RNA introduced into cells can selectively silence specific target genes.

Section 3 concludes the review of the analysis of exosomes (Chapter 11). RNA seems to be one of the main molecules transported in these organelles, and in the current literature, there is increasing evidence that confirms the important regulatory role performed by RNA contained in exosomes as effectors of various diseases. There are different types of extracellular vesicles—exosomes, microvesicles, and apoptotic bodies—which differ in their intercellular origin and their physicochemical characteristics and composition.

Exosomes are small particles (40–100 nm) generated by the majority of cell lines in culture as well as in vivo by cells forming tissues. They are released by all kinds of cells in physiological or pathological conditions, representing an important source of potential biomarkers. Its basic function is the intercellular communication by transferring proteins, RNAs, lipids, and DNA from one cell to another. Thus, this molecular information can change the physiology of the host cell at the transcriptional, post-transcriptional, or epigenetic level.

Exosomes have been involved in different biological processes, depending on cell origin, such as the regulation of immune response, antigen presentation, apoptosis, angiogenesis, inflammation, coagulation, and spread of neurodegenerative diseases, and in oncogenic processes. In Chapter 11, the currently employed methods for exosome isolation as well as for the purification of RNA molecules contained therein are reviewed.

Section 4. Cellular Models and Asthma

In Section 4, the main cell culture techniques applied to the study of asthma are reviewed. Cell cultures are an essential tool not only in basic research but also in the biotechnology industry and in the research of biomedicine. Several publications have reported studies in cultures of human bronchial epithelial cells, human T cells, or murine lung fibroblasts which analyze the immunological mechanisms that take place in the development and aggravation of respiratory and asthmatic diseases.

This knowledge has also been employed for new treatments and for therapeutic target development. As established in Chapter 12, the cell culture techniques allow obtaining information about in vitro models of study, which confers several advantages over animal experimentation.

The ability to control the environment in which they develop (as they allow a strict control of their physicochemical properties), the ability to obtain a large number of homogeneous cells (more complicated in animal experimentation), the opportunity to avoid the ethical conflicts of animal testing, and the economic aspects are some of the advantages over the animal models. However, some disadvantages may also be found, such as some instability if a large number of passes are made and the fact that cell cultures cannot always replace the in vivo tests. In Chapter 12, some of the most used cell culture techniques are described.

Chapter 13 discusses a protocol for transitional transfection in cell culture. Transfection is defined as the process whereby exogenous genetic material is introduced into cultured eukaryotic cells by using nonviral mechanisms. This technique is employed to evaluate the function of a specific gene into a specific environment, that is, the cell type employed. To this end, the model of study (cell type) should be carefully selected. In addition, an appropriate construction that includes the gene of interest inserted into an expression vector should be designed.

Variants of this method can be used with different applications; for instance, different transfections can be performed with constructions carrying genetic variants of the same gene to evaluate their effect over the expression. The effect over the protein expression can also be evaluated by collecting cell extracts and performing protein determinations by different techniques.

The cell line selected in the protocol of Chapter 13 is A549, a cell line derived from alveolar epithelium of human lung. Among the different mechanisms of transfection, the lipofection is one of the most commonly used. In addition, it is one of the easiest to use. In Chapter 13, different transfection techniques as well as requirements and recommendations for the selection of the expression vector are described. Eventually, the main transfection applications are mentioned at the end of the chapter.

Finally, Chapter 14 describes the protocol for the luciferase technique as a reporter system to measure the transcription activity of a target promoter that is introduced by transfection into a cell type. The promoter to be evaluated is inserted before the sequence of the luciferase gene into an expression vector host. After transfection, the chemiluminescence is detected by a luminometer, and the corresponding expression level is assigned. The method can be used with different promoter variants to evaluate their effects and to determine the specific promoter regions for gene regulation. Different applications and recommendations are also described in Chapter 14.

Section 5. Animal Models of Asthma and Pharmacological Applications

Asthma as a complex and heterogeneous disease is characterized by phenotypic diversity. Recent studies highlight the need to perform genetic and functional assays on each of the different phenotypes of asthma. To this end, animal models are being used to study the pathophysiological mechanisms involved in the different phenotypes of asthma and to identify the potential therapeutic targets for developing new drug treatments.

In Chapter 15, the application of mouse models to the study of asthma is reviewed. The mechanism of asthma in the mouse is also described, as well as the reasons why the use of mice offers several advantages over the use of other animal species. The scientific and economic reasons to consider the mouse an ideal model are described, in addition to the limitations of this model. Several clinical aspects are also discussed, such as how to generate mouse models that express chronic asthma or asthma in early life. Finally, technical aspects are also reviewed such as the selection of the proper adjuvant, mouse strain, allergen for sensitization, pathway of administration, time of induction, and so on.

In Chapter 16, different protocols to develop mouse models of asthma are provided. As previously described, the complexity of asthma research requires the development of appropriate models according to the different phenotypes of asthma. Finally, Chapter 17 describes how asthma mouse models are employed for pharmacological treatment investigation. Due to the high number of pharmacological compounds, there is an increasing need for methods that allow their preclinical evaluation. In this sense, testing in mouse models is an interesting approach. Chapter 17 describes different protocols for testing the effect of several compounds in different asthma models.

Section 6. Pharmacogenomics and Asthma

The last section of this book is devoted to the description of the pharmacogenomics studies performed in asthma. Pharmacogenomics is defined as the study of the role of genes in the drug response. The term pharmacogenomics refers to studies at a genome-wide level, and pharmacogenetics refers to studies of single genes in drug response. The main approach to determine the specific genetic condition of each individual is the study of the gene variants or the gene expression levels. At the pharmacological level, several factors should be taken into consideration such as drug absorption, distribution, metabolism, and elimination as pharmacokinetic factors and drug receptors and channels as pharmacodynamic factors.

Contrary to the traditional "one-dose-fits-all" and to the method of prescribing by "trial and error," pharmacogenomics is oriented to the development of a "personalized medicine," in which the genetic condition of each patient is studied and the therapy is optimized prior to application. This new point of view assumes an initial investment to determine the genetic background of each patient. However, once the genetic analysis is done, the costs dramatically decrease and are clearly rewarded from the economic point of view with the savings obtained by increasing treatment efficiency and decreasing toxicity and adverse drug reactions. In addition, the saving in the time employed by the medical personnel and especially the large improvement achieved in patient's quality of life should also be considered. This new way of medicine is being already employed in some cases, and it is foreseen that it will extend to all fields of medicine in the near future.

In Chapter 18, the main pharmacogenomics studies performed in asthma are reviewed. As has been mentioned, asthma is a good model of disease for these kinds of studies, due to the high variability between patient responses and available treatments. In this chapter, pharmacogenetic studies over the main asthma treatments, including beta-agonists, inhaled and systemic corticosteroids, leukotriene modifiers, anticholinergics, and biological drugs, are described.

Salamanca, Spain

María Isidoro-García
Catalina S. Sanz-Lozano
Ignacio Davila

Contents

Contributors

SARA CIRIA ABAD • *Department of Molecular Biology, Centro de Análisis Genéticos (Citogen), Zaragoza, Spain*

BELÉN GARCÍA-BERROCAL • *Department of Clinical Biochemistry, University Hospital of Salamanca, Salamanca, Spain; Salamanca Institute for Biomedical Research (IBSAL), University Hospital of Salamanca, Salamanca, Spain*

ASUNCIÓN GARCÍA-SÁNCHEZ • *Salamanca Institute for Biomedical Research (IBSAL), University Hospital of Salamanca, Salamanca, Spain; Department of Biomedical and Diagnostic Sciences, University of Salamanca, Salamanca, Spain*

VIRGINIA GARCÍA-SOLAESA • *Salamanca Institute for Biomedical Research (IBSAL), Salamanca, Spain; Department of Clinical Genetics, University Hospital of Navarra, Pamplona, Navarra, Spain*

IGNACIO DAVILA • *Department of Allergy, University Hospital of Salamanca, Salamanca, Spain; Department of Biomedical Science and Diagnosis, University of Salamanca, Salamanca, Spain; Salamanca Institute for Biomedical Research (IBSAL), Salamanca, Spain*

MARÍA ISIDORO-GARCÍA • *Department of Clinical Biochemistry, University Hospital of Salamanca, Salamanca, Spain; Department of Medicine, University of Salamanca, Salamanca, Spain; Salamanca Institute for Biomedical Research (IBSAL), Salamanca, Spain*

CATALINA S. SANZ-LOZANO • *Department of Microbiology and Genetics, University of Salamanca, Salamanca, Spain; Salamanca Institute for Biomedical Research (IBSAL), Salamanca, Spain*

ELENA MARCOS-VADILLO • *Department of Clinical Biochemistry, University Hospital of Salamanca, Salamanca, Spain; Salamanca Institute for Biomedical Research (IBSAL), Salamanca, Spain*

FERNANDO MARQUÉS-GARCÍA • *Salamanca Institute for Biomedical Research (IBSAL), Salamanca, Spain; Department of Clinical Biochemistry, University Hospital of Salamanca, Salamanca, Spain*

ALMUDENA SÁNCHEZ-MARTÍN • *Department of Pharmacy, University Hospital of Salamanca, Salamanca, Spain; Salamanca Institute for Biomedical Research (IBSAL), Salamanca, Spain*

IGNACIO SAN SEGUNDO-VAL • *Department of Clinical Biochemistry, University Hospital of Salamanca, Salamanca, Spain; Salamanca Institute for Biomedical Research (IBSAL), Salamanca, Spain*

Chapter 1

Applications of Molecular Genetics to the Study of Asthma

Catalina S. Sanz-Lozano, Virginia García-Solaesa, Ignacio Davila, and María Isidoro-García

Abstract

Asthma is a multifactorial disease. This fact, associated to the diversity of asthma phenotypes, has made difficult to obtain a clear pattern of inheritance. With the huge development of molecular genetics technologies, candidate gene studies are giving way to different types of studies from the genomic point of view.

These approaches are allowing the identification of several genes associated with asthma. However, in these studies, there are some conflicting results between different populations and there is still a lack of knowledge about the actual influence of the gene variants. Some confounding factors are, among others, the inappropriate sample size, population stratification, differences in the classification of the phenotypes, or inadequate coverage of the genes.

To confirm the real effect of the reported associations, it is necessary to consider both the genetic and environmental factors and perform functional studies that explain the molecular mechanisms mediating between the emergence of gene variants and the development of the disease.

The development of experimental techniques opens a new horizon that allows the identification of major genetic factors of susceptibility to asthma. The resulting classification of the population groups based on their genetic characteristics, will allow the application of specific and highly efficient treatments.

Key words Asthma, Candidate genes, GWAS, Next-generation sequencing, Phenotype

1 Study of Asthma as a Genetic Disease

Asthma has increased worldwide during the last decades; it is the most common chronic disease in children, with significant increases in economic and social costs [1]. Asthma is considered a complex disease caused by multiple factors (genetic, environmental, and epigenetic), with a polygenic character and phenotypically heterogeneous. The genetic approach to the study of complex diseases is focused on the identification of risk factors influencing the development of these diseases.

Different methodological approaches have been applied in order to understand the genetics factors underlying asthma. These studies are changing fast in last years; from the initial candidate genes studies to the high throughput techniques such as the

María Isidoro-García (ed.), *Molecular Genetics of Asthma*, Methods in Molecular Biology, vol. 1434,
DOI 10.1007/978-1-4939-3652-6_1, © Springer Science+Business Media New York 2016

Genome Wide Association Studies (GWAS) and the Next-Generation Sequencing (NGS). Although the emergence of the new technologies has allowed a changing in the methodological approach, the object of study has been the identification of gene variants associated to an increased disease risk.

The single nucleotide polymorphisms (SNPs) are single base-pair changes in the DNA sequence and appear at a frequency over 1 % in the population. Most of them have a high frequency in the human genome and a minimal impact on disease. Nevertheless, depending on their location in the gene, some SNPs have functional consequences such as changes in the function of the coded protein or in the gene expression levels. In contrast to the common SNPs, there are some genetics variants related to rare genetic disorders that appear at a very low frequency in populations.

The hypothesis of "common disease/common variant" (CD/CV) proposes that CDs have a different underlying genetic architecture than the rare disorders [2]. This interpretation arose from the observation that several complex diseases had multiple associations with many gene variants, which, separately, none seemed to have a definite effect.

Considering that the CVs have low penetrance but show heritability it seems that common genetic disorders must be determined mainly by multiple common SNPs or multiple common genetic factors, along with the environmental factors. However, in contrast to this idea, it is now believed that uncommon variants could have a greater contribution to the risk. In any case, the final identification of the full set of factors will enable the full understanding of mechanisms that trigger complex diseases as asthma, and thus which will enable the development of new forms of treatment.

2 Association Studies of Candidate Genes

Classically, the origins of the genetic studies in asthma started with the studies in families and twins that confirmed the genetic component of asthma and determined that asthma does not follow a Mendelian pattern but a polygenic inheritance. Linkage studies identified specific chromosomic regions related to asthma.

The Candidate Gene approach is based in an initial selection of the genes that are going to be studied. It requires a prior knowledge of the position (positional candidate) or function (functional candidate) of the gene. These approaches are based on a priori hypotheses and therefore have the advantage that their results are easily interpretable.

The disadvantage is that they are biased toward the specific mechanisms in which these genes are involved, and therefore do not allow advancing in the discovery of new genetic factors.

Another major drawback detected in these studies of candidate genes is the lack of replication between different reported associations. This may be due among others to ethnic factors, either by being developed in populations of different origin or because of the stratification of the populations, which hinders analysis.

Other reasons are differences in the phenotypic characterization of the groups included in the study so; it is essential implementing strict inclusion and exclusion criteria. In addition, a rigorous control of laboratory techniques is required. Finally, for the analysis and interpretation of the data is necessary controlling the statistical power as well as the possibility of issuing false positive results. Therefore, we must be very cautious with the statistical associations that have not been replicated in several studies.

Although initially some association studies were focused on the study of SNPs in a gene, it is more advisable to analyze multiple gene variants in the same locus (haplotypes and diplotypes). This approach is far more effective in avoiding spurious results. For the choice of the variants of interest, there are freely available platforms that provide information about the different variants located in a gene region [3, 4].

3 Genome Wide Association Studies

Studies of whole genome association, commonly called GWAS, are based on the simultaneous analysis of hundreds of thousands of gene markers. This has been possible thanks to the enormous progress in genotyping techniques that provide the high performance platforms, called microarrays. The GWAS analyzes the frequency of gene variants across the whole human genome and the results are used to predict the risk factors to disease development.

The GWAS studies are not subject to a prior hypothesis and allow identifying new genes and new pathogenic pathways. In contrast to linkage studies, they have a high resolution and detect risk variants with moderate effect. The number of polymorphisms analyzed per array has dramatically increased in the past few years, from about 10,000 to more than 1 million.

Currently, Affymetrix and Illumina provide important platforms used in GWAS analysis. The first commercially arrays were designed by Affymetrix about 15 years ago [5] and allowed genotyping 1494 SNPs per array. In the latest version (Affymetrix Genome Wide Human SNP Array 6.0) more than 906,600 SNPs and more than 946,000 CNV are analyzed. The Illumina platforms have also notably increased the number of SNPs per array in the latest decade with similar coverage and overall quality of data. Both platforms give an excellent genome-wide coverage for European and even for African populations. It should be noted that there are significant gaps in both [6].

With the increase in the number of polymorphisms per array, the complexity in the interpretation of the huge amount of generated information increases.

This complexity requires specialized professionals for its analysis, which further increases the costs of the process. These studies require strict statistical controls because of the large sample size used. The problem also arises from the sample characterization: if samples are not selected with strict inclusion criteria, large sample sizes are required to detect associations. Increasing the sample size favors the appearance of spurious associations, which necessitates the application of extreme significance thresholds for considering that these associations are valid.

The high initial cost of these first GWAS made the laboratories to pool samples to reduce costs. This approach allowed detecting some rare variants but offered no information on individual genotyping. As the density of SNPs on arrays has increased and the costs have been reduced, the microarrays are increasingly used for individual samples. This approach is much more powerful and allows the analysis of multiple phenotypes and endophenotypes at once, as well as the analysis of quantitative traits.

4 GWAS Results in Asthma

In the last decades, more than 30 studies of GWAS have been published on asthma and have identified new and interesting genes confirming the role of some functionally relevant genes previously described (*see* Table 1). Although most GWAS data are referred to European populations some differences regarding ethnicity have been observed. So far, the heritability of asthma has not been fully elucidated and this remarks the disease heterogeneity.

These results also suggest the importance of the phenotypes in asthma, which remarks the need to better define the disease and to improve its phenotyping in the GWAS analysis. For instance, the recent results differ between the genes involved in the development of asthma and the genes involved in the severity of the disease. Descriptive variables, such as age of onset, trigger factors, frequency, and severity of symptoms and the response to treatment are often used to classify the different phenotypes of asthma. Thus asthma is not considered a single disease but a conjunction of different diseases or different phenotypes.

The discovery of accurate markers would be helpful in dissecting the different entities of the syndrome of asthma and would assist in the disease prevention and treatment at the individual level. To this end, some identifiers reported in phenotypes of asthma are among others asthma onset in childhood, asthma onset in adulthood, occupational asthma, asthma difficult to treat (persistent severe), atopic asthma, nonatopic asthma, lung function, blood eosinophil count, etc.

Table 1
GWAS and asthma phenotypes

Asthma subphenotype	Gene	References
YKL-40 levels	*CHI3L1*	[34]
Total serum IgE	*FCER1A* *RAD50* *STAT6*	[24]
Eosinophil count	*GATA2* *GRFA2* *IKZF2* *IL1RL1* *IL5* *IL33* *MYB* *SH2B3* *WDR36*	[28]
Eosinophilic esophagitis	*TSLP* *WDR39* *13q31.1 region*	[29]
Atopic dermatitis	*FLG* *C11orf30*	[30]
Lung function	*GSTCD* *TNS1* *HHIP* *HTR4* *AGER* *THSD4*	[19]
	GRP126 *ADAM198* *AGER/PPT2* *FAM13A* *PID1* *HTR4* *INTS12/GSTCD* *NPNT*	[20]
Childhood onset of asthma	*ORMDL3* *GSDMB*	[7]
	CRB1 *DENND1B*	[16]
Adulthood onset of asthma	*IL18R1* *HLA-DRB1* *HLA-DQ* *IL33* *ORMDL3* *GSDML* *SMAD3* *IL2RB* *SLCA22A5* *IL13* *RORA*	[17]
Severe difficult to treat asthma	*RAD50* *IL13* *HLA-DR_DQ*	[24]

5 GWAS and Asthma Phenotypes

Regarding the early onset asthma, the first GWAS that reported on childhood asthma was performed in 2007 by the GABRIEL consortium, Phase I [7]. In this study, the chromosomal region 17q21 was related with susceptibility to childhood asthma. Inside this region is located the *ORMDL3* gene (OMR-like protein 3). Variants in this gene were associated and correlated with differences in mRNA levels of *ORMDL3*, pointing to a functional role in gene expression. Different studies with several ethnically diverse populations confirmed this association [8–10].

The putative role of ORMDL3 remains unknown, with some studies pointing to intracellular calcium homeostasis, which would induce inflammation mechanisms [11], and other to a role in regulating protein metabolism of sphingolipids [12]. Some gene variants that are located in this chromosomal region have also been associated to asthma such as of *GSDMB* gene that regulates apoptosis of gastric epithelium, so it is thought that more than one gene in this region 17q21 may contribute to asthma [13].

It has been seen an extensive linkage disequilibrium between *ORMDL3* and *GSDMB* genes what along with coregulation of their expression makes it hard to define the real associations of these genes. Interestingly these associations were replicated in different childhood populations. The region 17q21 has also been associated with other inflammatory diseases such as inflammatory bowel disease and diabetes, what could also hint at a more general or basic role of this region in chronic inflammatory conditions [14, 15].

Sleiman et al. reported another region related with childhood asthma in a population of children with persistent asthma that required daily inhaled glucocorticoid therapy. A novel asthma-susceptibility locus containing the *CRB1* and *DENND1B* genes was identified in chromosome 1q31 [16].

Regarding asthma onset in adulthood, 1947 patients with adult-onset asthma were genotyped on the phase II GWAS of the GABRIEL consortium. Among the associations, *HLA-DQ* genes were seen to have a clear association with the onset of asthma in adulthood [17]. In 2011, a GWAS on adult asthma in a Japanese population reported five candidate loci [18]. The most significant association with adult asthma was observed in rs404860 in the major histocompatibility complex (MHC) region on chromosome 6p21, which is close to rs2070600 that was previously reported for association with FEV1/FVC [19, 20].

Human leukocyte antigen studies conducted in European workers have defined alleles of the major histocompatibility complex class II and haplotypes associated with diisocyanate asthma. Recently, certain genotypes coding for glutathione-S-transferase and N-acetyltransferase were associated with slow acetylation phenotypes in diisocyanate asthma. In addition, combinations of *IL4R* and *CD14* SNPs were significantly associated with diisocyanate asthma, but only in workers exposed to hexamethylene diisocyanate. A recent GWAS conducted in Korea identified several SNPs of the α-T-catenin coding gene that were significantly associated with diisocyanate asthma [21].

A recent GWAS identified *PYHIN1* as a novel gene associated to asthma in individuals of African descent [22]. Variants on *IL6R* and in *GARP* genes have also been reported as significant associated in a GWAS performed on an Australian population [23]. Other genes associated with asthma through a large scale GWAS were *IL1RL1*, *IL18R1*, *HLA-DQ*, *IL33*, *SMAD3*, *ORMDL3*, *GSDMB*, and *IL2RB* [17].

Regarding severe asthma, there have been reported associations with the *RAD50-IL13* region on chromosome 5q31.1 and *HLA-DR_DQ* region on chromosome 6p21.3 [24]. The atopic march theory establishes that atopy is a precursor for the future development of asthma, although genetics results, regarding atopy and asthma, point towards different backgrounds for these two entities. A GWAS performed in a German population of 1530 subjects identified some associated variants in genes *FCER1A* (High affinity IgE receptor), *STAT6* (a transcription factor), and *RAD50* (a DNA repair protein) regarding IgE serum levels [25]. In this study, some variants on the 5q31 region were associated with IgE. These regions have also been associated with severe asthma [24], pointing to a putative role in atopic asthma. Several genes coding cytokines are located in this region as *IL3*, *IL4*, *IL5*, *IL13*, and *GM-CSF*.

In contrast to all the genes found to be associated with asthma in the GWAS study of the GABRIEL consortium, only *HLA-DRB1* was significantly associated with serum IgE levels [17], what again suggests a different genetic background for atopy and IgE mediated asthma and the phenotypes of asthma. While it is still unclear whether IgE is a secondary event in the development of the asthmatic phenotype, it seems that atopic and nonatopic asthma could be two diverse entities in terms of genetic susceptibility [26]. Skin prick tests (SPT) and allergen-specific IgE levels have also been analyzed as genetic susceptibility signals in a recent GWAS but no consistent patterns of association were found [27].

In another study performed mainly in adult populations, sensitization to grass was associated with the *HLA-DRB4* locus. Taken together, it could be hypothesized that genetic determinants for atopy phenotypes and for specific sensitization differ [26].

The eosinophil count has also been considered in GWAS of asthma, taking into account the important role of eosinophils in asthma. *IL1RL1*, *IKZF2*, *GATA2*, *IL5*, *SH2B3*, *MYB*, *WDR36*, *IL33*, and *GRFA2* genes were associated in a population of 9393 subjects, including, European, Asian, and American ethnicities [28]. A GWAS study on eosinophilic esophagitis (EoE) over a population of 181 cases and 1974 controls of European and American ethnicity reported associations with, *TSLP*, *WDR39*, and 13q31.1 regions [29].

Atopic dermatitis could be considered as a phenotype associated to asthma, since both diseases are frequently presented together. The first GWAS of atopic dermatitis was performed on a German population and showed associations for the filaggrin (*FLG*) gene and for the C11orf30 locus located on 11q13.5 chromosome where the *LRRC32* gene is located [30]. In another GWAS on a Chinese population, the *TMEM232* and *SLC25A46* genes were associated with atopic dermatitis [31]. Then a meta-analysis on atopic dermatitis performed in a European population revealed associations for *OVOL1*, *ACTl9*, *KIF3A*, *IL4*, and *IL13* genes [32]. Finally, a Japanese study found associations for *IL1RL1/IL18R1/IL18RAP*, *GLB*, *CCDC80*, the *MHC* region, *CARD11*, *ZNF365/EGR2*, *OR10A3/NLRP10*, and *CYP24A1/PFDN4* genes [33].

The *YKL-40* level has also been taken into consideration in GWAS analysis since it is increased in asthmatic patients. The *CHI3L1* gene has been found associated with the *YKL-40* serum levels in a population of European and American ethnicity [34]. SNPs in this gene have also been associated with lung function and bronchial hyperresponsiveness.

Another asthma characteristic is the reduced pulmonary function. Reduction of FEV1/FVC is a general characteristic of obstructive lung diseases so some GWAS have studied its genetic relationship with asthma. A GWAS on a European and American population of 7691 subjects found an association of lung function measure by FEV1/FVC ratio with the 4q31 region that is close to the *HHIP* gene [35]. *HHIP* is part of the hedgehog-signaling pathway and may play a role in embryonic lung development [36].

A meta-analysis of the SpiroMeta consortium with 20,288 European individuals revealed an association of lung function with *GSTCD*, *TNS1*, 4q31 region (near *HHIP*), *HTR4*, *AGER*, and *THSD4* genes [19]. Finally a CHARGE consortium meta-analysis in 20,890 European and American subjects showed associations of lung function with *HHIP*, *GRP126*, *ADAM198*, *AGER/PPT2*, *FAM13A*, *PID1*, *HTR4*, *INTS12/GSTCD*, and *NPNT* genes [20]. These associations of genes with lung function measurements and asthma may help to identify lung specific mechanisms in asthma development in contrast to general inflammatory mechanisms (such as eosinophilia) that also play a role in asthma development [26].

6 Final Considerations over GWAS Results in Asthma

The results obtained in GWASs conducted in the last decade, allow us to draw certain conclusions that may be important when considering future studies. Some points to consider are the appropriate clinical characterization of patients, taking special care in selecting specific phenotypes of asthma, since different groups seem to have different genetic determinants.

Equally important is the selection of a population of controls strictly characterized, being this one of the most difficult populations to achieve. Another major problem of these studies has been carelessness in the use of quality measures in experimental development. The huge amount of results obtained in these studies makes necessary the bioinformatics and biostatistical advice for employing the appropriate tests, the adequate sample sizes, and avoid false positive results due to the development of multiple comparisons. Finally, it is necessary to validate the GWAS results with specific gene association studies, and with functional analysis of genes for which these associations were detected.

7 Next-Generation Sequencing

Fred Sanger and colleagues published in 1977 the discovery of a new technique that allowed determining the sequence of a DNA fragment [37]. This discovery was an essential tool for genetic studies conducted in the following decades in which many genes were analyzed. This technology was applied even for the study of the whole human genome being one of the main bases for its determination. This technique began by sequencing small fragments of about 25 base pairs (bp) reaching now up a length of about 750 bp.

Although the size of the sequenced fragments has markedly increased, the difficulty of analyzing multiple fragments in parallel remains a problem. Even after the methodological improvements application as the use of capillary electrophoresis, the amount of reactions simultaneously sequenced and their high costs continued to pose a problem for the scientific community that demands a more efficient and economical method.

While the Sanger method of automatic sequencing is considered a first-generation technology, new methods are named next-generation sequencing (NGS). NGS technologies allow the sequencing of the entire genome of an individual in several hours. Unlike GWASs, NGS shows the complete genome information of each individual, making possible the detection of rare gene variants [38]. The main disadvantages are that they require advanced bioinformatic treatments and their high costs.

Regarding the immobilization of the fragments there are alternatives of choice. First, there are solid supports where primer molecules are covalently attached and uniformly distributed; subsequently, the template molecules join to these primers through common adapters. A second option consists in joining the single chain fragments to the solid surface. Eventually, the polymerase anchored to the support to which a primed template is bound can be used.

The next step is the sequencing process that also presents several alternatives. The Cyclic Reversible Termination (CRT) method is based on the incorporation of a fluorescent modified nucleotide that binds to the primed template by the DNA polymerase. Then the identity of the incorporated nucleotide is determined depending on the kind of fluorescence generated. Finally the terminator/inhibitor group and the fluorescent label are cut [39]. Another method is the Sequencing By Ligation where primed template fragments are hybridized with fluorescent-labeled complementary molecules and the binding is sealed by the DNA ligase [40]. The Single-Nucleotide addition pyrosequencing method is based on measuring the pyrophosphate emission through its conversion into light [41]. The identification of the nucleotide incorporated at each position is determined by the type of light peak generated [42].

Finally, the fragments sequenced must be assembled and aligned. To do this database, reference sequences can be used; it is also possible to perform a "de novo" assembly. In the vast majority of cases, the alignments to the reference sequences are conducted; however, this method is not always possible and has some disadvantages, such as problems with the assembly generated for highly repeated regions. "De novo" assemblies have been employed mainly with bacterial genomes but are not frequent in humans.

In 2005, the first NGS technology was released, based in the pyrosequencing method [43]. The appearance of this first NGS system allowed analyzing 50 times more sequences, with 6 times lower costs than the previous systems. This platform employs DNA amplified samples by the emPCR system on which a pyrosequencing process is subsequently performed. The main problem described for this platform is the identification of insertions and deletions, especially in homopolymeric regions.

One year later, a second platform was commercialized [44], becoming one of most frequently used platforms. The method employed for fragments amplification is the solid-phase sequencing followed by the CRT. The main problem described for the results generated with this platform is the detection of substitutions mainly when the previous nucleotide is a G [45]. In addition, it has been described a poor representation of sequences rich in AT or GC [45–47].

Other technology released was the Sequencing by Oligo Ligation Detection in 2007 [48]. The platform employs fragmented and amplified DNA samples by emPCR on which sequencing is performed by ligation method. The main problem described for this platform is the substitution and the low representation of sequences rich in AT or in GC [47].

Besides focusing on methods for analyzing DNA variation, NGS technologies have also been applied to the analysis of RNA expression (RNAseq) [49–51], immunoprecipitation-based protein–DNA or protein–RNA interaction mapping, and DNA methylation using bisulfite-mediated cytosine conversion [48, 52]. When the cost of sequencing a human genome approaches $1000, the NGS will probably replace the genotyping by GWAS.

One very important issue regarding the NGS results is the validation. By now data from this technology should be confirmed by a gold standard methodology. For this purpose, Sanger technique is been used to confirm the results. In addition, the big amount of data obtained is difficult to manage and, by now, scarce biological and clinical validation exists.

Despite having identified numerous disease susceptibility genes, the results obtained to date with all the platforms described here have not managed to provide a full explanation to the asthma heritability and most variants found have little effect. Some genetic factors that still are not taken into account could explain this missing heritability such as variants (rare and common), copy number variations, or epigenetic factors.

Acknowledgment

This work was supported by a grant of the Junta de Castilla y León ref BIO/SA73/15.

References

1. Call for Global Action on Chronic Respiratory Diseases organized by the World Health Organization and the European Federation of Allergy and Airways Diseases (Rome, 11 June 2009). http://www.efanet.org/enews/press.html/, http://www.who.int/respiratory/gard/en/
2. Mayo O (2007) The rise and fall of the common disease-common variant (CD-CV) hypothesis: how the sickle cell disease paradigm led us all astray (or did it?). Twin Res Hum Genet 10(6):793–804
3. International HapMap Consortium, Frazer KA, Ballinger DG, Cox DR et al (2007) A second generation human haplotype map of over 3.1 million SNPs. Nature 449(7164):851–861
4. Barrett JC, Fry B, Maller J et al (2005) Haploview: analysis and visualization of LD and haplotype maps. Bioinformatics 15(2):263–265
5. Wang DG, Fan JB, Siao CJ et al (1998) Large-scale identification, mapping, and genotyping of single-nucleotide polymorphisms in the human genome. Science 280:1077–1082
6. Saccone SF, Bierut LJ, Chesler EJ et al (2009) Supplementing high-density SNP microarrays for additional coverage of disease-related genes: addiction as a paradigm. PLoS One 4(4):e5225. doi:10.1371/journal.pone.0005225
7. Moffatt MF, Kabesch M, Liang L et al (2007) Genetic variants regulating ORMDL3 expression contribute to the risk of childhood asthma. Nature 448(7152):470–473

8. Galanter J, Choudhry S, Eng C et al (2008) ORMDL3 gene is associated with asthma in three ethnically diverse populations. Am J Respir Crit Care Med 177(11):1194–1200
9. Hirota T, Harada M, Sakashita M et al (2008) Genetic polymorphism regulating ORM1-like 3 (*Saccharomyces cerevisiae*) expression is associated with childhood atopic asthma in a Japanese population. J Allergy Clin Immunol 121(3):769–770
10. Leung TF, Sy HY, Ng MC et al (2009) Asthma and atopy are associated with chromosome 17q21 markers in Chinese children. Allergy 64(4):621–628
11. Cantero-Recasens G, Fandos C, Rubio-Moscardo F et al (2010) The asthma-associated ORMDL3 gene product regulates endoplasmic reticulum-mediated calcium signaling and cellular stress. Hum Mol Genet 19(1):111–121
12. Breslow DK, Collins SR, Bodenmiller B et al (2010) Orm family proteins mediate sphingolipid homeostasis. Nature 463(7284):1048–1053
13. Verlaan DJ, Berlivet S, Hunninghake GM et al (2009) Allele-specific chromatin remodeling in the ZPBP2/GSDMB/ORMDL3 locus associated with the risk of asthma and autoimmune disease. Am J Hum Genet 85(3):377–393
14. McGovern DP, Gardet A, Törkvist L et al (2010) Genome-wide association identifies multiple ulcerative colitis susceptibility loci. Nat Genet 42(4):332–337
15. Barrett JC, Clayton DG, Concannon P et al (2009) Genome-wide association study and meta-analysis find that over 40 loci affect risk of type 1 diabetes. Nat Genet 41(6):703–707
16. Sleiman PM, Flory J, Imielinski M et al (2010) Variants of DENND1B associated with asthma in children. N Engl J Med 362:36–44
17. Moffatt MF, Gut IG, Demenais F et al (2010) A large-scale, consortium-based genome wide association study of asthma. N Engl J Med 363(13):1211–1221
18. Hirota T, Takahashi A, Kubo M et al (2011) Genome-wide association study identifies three new susceptibility loci for adult asthma in the Japanese population. Nat Genet 43:893–896
19. Repapi E, Sayers I, Wain LV et al (2010) Genome-wide association study identifies five loci associated with lung function. Nat Genet 42:36–44
20. Hancock DB, Eijgelsheim M, Wilk JB et al (2010) Meta-analyses of genome-wide association studies identify multiple loci associated with pulmonary function. Nat Genet 42:45–52
21. Bernstein DI (2011) Genetics of occupational asthma. Curr Opin Allergy Clin Immunol 11(2):86–89
22. Torgerson DG, Ampleford EJ, Chiu GY et al (2011) Meta analysis of genome-wide association studies of asthma in ethnically diverse North American populations. Nat Genet 43:887–892
23. Ferreira MA, Matheson MC, Duffy DL et al (2011) Identification of IL6R and chromosome 11q13.5 as risk loci for asthma. Lancet 378:1006–1014
24. Li X, Howard TD, Zheng SL et al (2010) Genome-wide association study of asthma identifies RAD50-IL13 and HLA-DRDQ regions. J Allergy Clin Immunol 125: 328–335
25. Weidinger S, Gieger C, Rodriguez E et al (2008) Genome-wide scan on total serum IgE levels identifies FCER1A as novel susceptibility locus. PLoS Genet 4(8):e1000166
26. Binia A, Kabesch M (2012) Respiratory medicine – genetic base for allergy and asthma. Swiss Med Wkly 142:1–11
27. Wan YI, Strachan DP, Evans DM et al (2011) A genome-wide association study to identify genetic determinants of atopy in subjects from the United Kingdom. J Allergy Clin Immunol 127(1):223–231
28. Ramasamy A, Curjuric I, Coin LJ et al (2011) A genome-wide meta-analysis of genetic variants associated with allergic rhinitis and grass sensitization and their interaction with birth order. J Allergy Clin Immunol 128:996–1005
29. Rothenberg ME, Spergel JM, Sherrill JD et al (2010) Common variants at 5q22 associate with pediatric eosinophilic esophagitis. Nat Genet 42:289–291
30. Esparza-Gordillo J, Weidinger S, Fölster-Holst R et al (2009) A common variant on chromosome 11q13 is associated with atopic dermatitis. Nat Genet 41:596–601
31. Sun LD, Xiao FL, Li Y et al (2011) Genome-wide association study identifies two new susceptibility loci for atopic dermatitis in the Chinese Han population. Nat Genet 43: 690–694
32. Paternoster L, Standl M, Chen CM et al (2011) Meta-analysis of genome-wide association studies identifies three new risk loci for atopic dermatitis. Nat Genet 44:187–192
33. Hirota T, Takahashi A, Kubo M et al (2012) Genome-wide association study identifies eight new susceptibility loci for atopic dermatitis in the Japanese population. Nat Genet 44:1222–1226

34. Ober C, Tan Z, Sun Y et al (2008) Effect of variation in CHI3L1 on serum YKL-40 level, asthma risk, and lung function. N Engl J Med 358:1682–1691
35. Wilk JB, Chen TH, Gottlieb DJ et al (2009) A genome-wide association study of pulmonary function measures in the Framingham Heart Study. PLoS Genet 5:e1000429
36. Wilk JB, Chen TH, Gottlieb DJ et al (2009) A genome-wide association study of pulmonary function measures in the Framingham Heart Study. PLoS Genet 5(3):e1000429. doi:10.1371/journal.pgen.1000429
37. Sanger F, Air GM, Barrell BG et al (1977) Nucleotide sequence of bacteriophage phi X174 DNA. Nature 24:687–695
38. Buermans HP, den Dunnen JT (2014) Next generation sequencing technology: advances and applications. Biochim Biophys Acta 1842(10):1932–1941
39. Metzker ML (2005) Emerging technologies in DNAsequencing. Genome Res 15: 1767–1776
40. Tomkinson AE, Vijayakumar S, Pascal JM et al (2006) DNA ligases: structure, reaction mechanism, and function. Chem Rev 106:687–699
41. Ronaghi M, Uhlén M, Nyrén P (1998) A sequencing method based on real-time pyrophosphate. Science 281:363–365
42. Ronaghi M, Karamohamed S, Pettersson B et al (1996) Real-time DNA sequencing using detection of pyrophosphate release. Anal Biochem 242:84–89
43. Margulies M, Egholm M, Altman WE et al (2005) Genome sequencing in micro fabricated high-density picolitre reactors. Nature 437:376–380
44. Bentley DR, Balasubramanian S, Swerdlow HP et al (2008) Accurate whole human genome sequencing using reversible terminator chemistry. Nature 456:53–59
45. Dohm JC, Lottaz C, Borodina T et al (2008) Substantial biases in ultra-short read data sets fromhigh-throughput DNA sequencing. Nucleic Acids Res 36:e105
46. Hillier LW, Marth GT, Quinlan AR et al (2008) Whole-genome sequencing andvariant discovery in *C. elegans*. Nat Methods 5:83–188
47. Harismendy O, Ng PC, Strausberg RL et al (2009) Evaluation of next generationsequencing platforms for population targetedsequencing studies. Genome Biol 10(3):R32. doi:10.1186/gb-2009-10-3-r32, Epub 2009 Mar 27
48. Valouev A, Ichikawa J, Tonthat T et al (2008) A high-resolution, nucleosomeposition map of C. elegans reveals a lack of universalsequence-dictated positioning. Genome Res 18:1051–1063
49. Chen G, Wang C, Shi T (2011) Overview of available methods for diverse RNA-Seq data analyses. Sci China Life Sci 54(12):1121–1128
50. Qi YX, Liu YB, Rong WH (2011) RNA-Seq and its applications: a new technology for transcriptomics. Yi Chuan 33(11):1191–1202
51. Yick CY, Zwinderman AH, Kunst PW et al (2013) Transcriptome sequencing (RNA-Seq) of human endobronchial biopsies: asthma versus controls. Eur Respir J 42(3):662–670
52. Pascual M, Roa S, García-Sánchez A et al (2014) Genome-wide expression profiling of B lymphocytes reveals IL4R increase in allergic asthma. J Allergy Clin Immunol 134(4): 972–975

Chapter 2

Assessment and Validation of New Genetic Variants: A Systematic In Silico Approach

Belén García-Berrocal, Asunción García-Sánchez, and María Isidoro-García

Abstract

The application of new high-throughput technologies to the study of asthma and other complex diseases is providing a huge amount of genetic information. Particularly, next-generation sequencing generates thousands of variants that have not been previously related to the studied diseases. These new genetic variants require validation both at methodological and clinical level. In this sense, for methodological validation the capillary electrophoresis (CE) sequencing based on Sanger technique continues to be the reference technique. The aim of this chapter is to provide a systematic approach of the in silico procedures to the confirmation of the new genetic variants and to their assessment for the pathogenic condition determination.

Key words Asthma, In silico, Gene, Genotyping, Sanger, Polymorphisms, Sequencing

1 Introduction

The rapid development of technology provides new methodologies applied to the genotyping of different diseases. Asthma as a complex disease depends on genetic as well as epidemiological factors. The complex pattern of inheritance of asthma required the application of high-throughput technologies to the study of genetic background. In this sense next-generation sequencing arises as a very promising technique. However, as other "omics" methodologies, it generates thousands of data difficult to interpret.

Considering that millions of bases are analyzed, there is an important need to validate this huge amount of information. The validation has two different approaches, biological and methodological. Clinical validation of the huge amount of results requires many studies in patient populations to confirm the association of these variants with the disease. On the other hand, for the methodological validation, a reference technique as the capillary electrophoresis (CE) is needed.

María Isidoro-García (ed.), *Molecular Genetics of Asthma*, Methods in Molecular Biology, vol. 1434,
DOI 10.1007/978-1-4939-3652-6_2, © Springer Science+Business Media New York 2016

The continuous process of validation will increase the confidence of the scientific community in the technique as well as in the generated data. In this chapter, we present a protocol that includes an in silico approach to the confirmation of new genetic variants and to assess their pathogenicity.

This protocol is useful not only to identify and confirm genetic variants detected by other techniques as NGS but also to interrogate the segregation pattern of a punctual mutation in the family of the index case. In addition, when thousands of bases are analyzed, multiple technical problems can arise. One of them is the identification of false positives. The CE technique also allows discarding this spurious information.

In the validation of a new genetic variant, the complete sequence of the gene must be known to study and design the most appropriate primers for the PCR. This in silico protocol consists of two sections, the one previous to the laboratory technique and the other corresponding to the interpretation of data.

2 Materials

For this in silico protocol the materials are basically the following available databases:

USCS Genome Browser page Home: https://genome.ucsc.edu/.

PRIMER 3: http://biotools.umassmed.edu/bioapps/primer3_www.cgi.

Primer Beacon Designer: http://www.premierbiosoft.com/qpcr/index.html.

Reverse Complement program: http://www.bioinformatics.org/sms/rev_comp.html.

EditView Mac: http://www.appliedbiosystems.com.

Chromas PC: http://technelysium.com.au.

GeneBank: http://www.ncbi.nlm.nih.gov/genbank.

Human Genome Variation Society: www.hgvs.org/mutnomen.

The Human Gene Mutation Database: www.hgmd.cf.ac.uk/ac/index.php.

Clin Var: http://www.ncbi.nlm.nih.gov/clinvar/.

Poly-Phen (Polymorphism Phenotyping): http://genetics.bwh.harvard.edu/pph2/bgi.shtml.

The National Center for Biotechnology Information: http://www.ncbi.nlm.nih.gov/.

Serial Cloner: http://serialbasics.free.fr/Serial_Cloner.html.

SIFT (Sorting Intolerant From Tolerant): http://sift.bii.a-star.edu.sg/.

ALFRED (The Alelle Frecuency database): http://alfred.med.yale.edu/.

OMIM: https://omim.org/; http://www.ncbi.nlm.nih.gov/omim.

3 Methods

3.1 Identification of the Gene Sequence

Once a new genetic variant is identified by NGS, it is necessary to know the complete sequence of the gene (*see* **Note 1**). Different online platforms provide the entire sequence of genes; as an example, the USCS Genome Browser page can be used.

1. Select Genome Browser on the top of the left panel.
2. Select the following:
   - Group: Mammal.
   - Genome: Human.
   - Assembly: Latest version (*see* **Note 2**).
3. Search team: Enter the name of the Gene and click the button Submit.
4. Select the proper sequence at RefSeq Genes.
5. Click the highlighted gene, on the left column.
6. Select genomic sequence from assembly (Below links to sequence).
7. Choose from the boxes below, as convenience:
   - Promoter/Upstream (select the number of bases).
   - 5′UTR Exons.
   - CDS Exons (*see* **Note 3**).
   - 3′UTR Exons.
   - Introns.
   - Downstream (select the number of bases).
8. Click submit.
9. Find the region of interest within your sequence (*see* **Note 4**) and look for the specific variant.
10. If the variant has been previously described you can use NCBI Clin Var platform to identify the position.

3.2 Design of Primers for Polymerase Chain Reaction

The sequence and length of the primers are essential characteristics for obtaining an optimal PCR reaction. Different software can be used to design the best primers for any specific genomic region. As an example:

1. Open the web for PRIMER 3.
2. In the space provided, copy the sequence you want to analyze.
3. Introduce the proper information about the amplicon (product size) and the primer and click Primer.

   The following suggestions can be considered for proper design of the primers that are to be used:

(a) Optimal size of amplicons should be approximately 500 bp (base pairs) (*see* **Note 5**).

(b) Primer minimum size: 20 bp, optimal: 22 bp and maximum: 25 bp. The primers should have at least 18 bases to ensure that it hybridizes with the DNA sequence.

(c) Tm (melting temperature) 59–61, ideal is 60 °C, no more than 2° of difference between the couple. To modify the temperature remove or add one nucleotide in 5′, using the following formula to calculate the *Tm*:

$$Tm = 69.3 + (0.41 \times \text{GC}\,\%) - (650 / \text{primer length})$$

(*See* **Note 6**).

(d) The amount of GC bases must be between 45 and 55 %; it is widely accepted between 40 and 60 %. To get a Tm above the recommended, the primers with lower GC content must be longer.

(e) In principle, there is no minimum distance from the primer to the SNP but the best is that it is centered. Design the primer not immediately before the SNP. Remember to leave 50 bases immediately after, because the first sequence obtained can be quite vague.

(f) GC Clam: Is the presence of G or C in the last five bases of the 3′end. Helps stabilize the binding of the first to sequence. Avoid over 3 Gs or Cs.

(g) Avoid secondary structures:

- Harpings: Intermolecular interactions of the primer. Would be tolerated if presented in the 3′-end with a $\Delta G = -2$ kcal/mol or whether they are internal with a $\Delta G = -3$ kcal/mol.
- Self-dimers: Intermolecular interactions of the same primer. Tolerated if presented in the 3′-end with a $\Delta G = -5$ kcal/mol or whether they are internal with a $\Delta G = -6$ kcal/mol.
- Cross-dimers: Intermolecular interactions of the primers forward and reverse. Tolerated if presented in the 3′-end with a $\Delta G = -5$ kcal/mol or whether they are internal with a $\Delta G = -6$ kcal/mol.
- The ΔG o Gibbs free energy is the measure of the spontaneity of the reaction. The stability of the secondary structures, such as dimers or Harpings, is measured by its ΔG, i.e., the energy required to break them. The more negative, the greater the stability of the secondary structures.

(h) Runs: Repetitions of the same base within the primer. Only four are allowed.

(i) Avoid cross-homology: Primers designed for the specific sequence should not amplify other genes; you should make a BLAT and an in silico PCR (see below) to check it.

4. Keep the information about the design conditions of primers because it will be later needed for the PCR design.
5. Select the Primer Beacon Designer to analyze the stability:
6. Select Beacon Designer software Free, then the Launch Free Edition page designer opens (if new user you must register) after that Launch Beacon Designer, select the Oligo analysis, and Select SYBR Green
7. Paste the sequence of the primer forward (sense primer) and then the reverse (antisense primer), both 5′-3′.
8. Give a click to Analyze. A box with all indications Tm, GC%, GC Clamp, and the secondary structures for sense primer will appear.
9. Check the primers using the UCSC program Genome Browser Home. Rule out the presence of SNPs in the primers. To check if they stick to another part of the genome, click in the UCSC Genome Browser Home program; on the top bar select the Tool options BLAT.
10. Skip primers separately analyzed.
11. Confirm that the % of correspondence is 100 % and that is the correct chromosome.
12. Locate the primers in the gene sequence (*see* **Note 7**).

Now you have designed the primers and the conditions for PCR and the subsequent capillary electrophoresis, proceed with the laboratory techniques.

3.3 Analysis of Results

The sequencers most frequently used provide the results in ABI format that are compatible with the common available programs.

Once we get the chromatogram of the nucleotide sequence, this process automatically makes programs that read the chromatograms but it should be manually checked because sometimes it fails to assign the bases.

1. Open the program Edit view or Chromas.
2. Open your sequence.
3. Discard the beginning and the end of the sequences that are vague and confusing.
4. Check the background noise (*see* **Note 8**).
5. Review the entire fragment looking for overlapping sequences. When there are insertions or deletions in the amplicon, the sequence can begin normally but there is a double reading from a certain point (*see* **Note 9**).

6. Review the entire sequence looking for N (bases which the software is unable to correctly interpret). Assign the combination of bases according to Table 1 (*see* **Note 10**).

   For the wild-type homozygous there is no change with respect to the original base; for the heterozygote there are two overlapping bases, one of each allele, and it will appear as an N. For the homozygous mutated, the base is different in the two alleles; there will be only one base but different from the original sequence (Fig. 1).

7. You can also align the sequence. The Basic Local Alignment Search Tool (BLAST) tool is a search engine that uses a heuristic algorithm (Smith-Waterman) to performing sequence alignment. This tool allows you to compare a model sequence (query) with all sequences stored in the database. The program searches in the database and submits those sequences most similar to the model sequence.

It can be done from the web site of GenBank or from UCSC Genome Browser.

The correct nomenclature for a mutation can be found on the Internet Human Genome Variation Society page.

Table 1
Encoding for the four bases (A, C, T, G) and for ambiguous positions in DNA sequence

Code	Meaning
A	A
T/U	T
C	C
G	G
M	A or C
K	G or T
R	A or G
Y	C or T
S	C or G
W	A or T
H	A or C or T
V	A or C or G
B	C or G or T
D	A or G or T
X/N	G or A or T or C

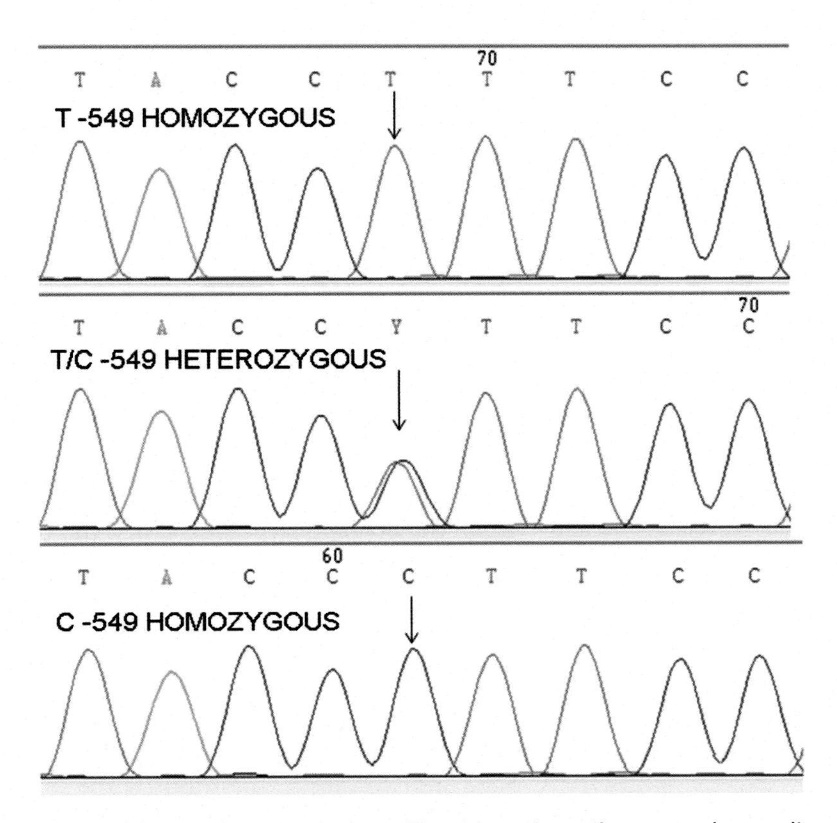

Fig. 1 DNA sequence chromatograms. The picture shows the sequencing results of three PCR amplicons from genomic DNA showing a polymorphic position. Heterozygous single-nucleotide polymorphism (SNP) appears as two peaks of different colors instead of just one, showing both nucleotides simultaneously. One allele carries a T, and the other has a C. Both peaks are present, but at roughly half the height they would show that they were homozygous. Note that the software may list that base position as an "N"

3.4 Assessment of New Variants

Once defined and confirmed the change in the DNA sequence, it is necessary assessing the clinical significance. There are some initiatives that provide guidelines for the assessment of the variants although there is still lack of consensus that can be easily adopted [1, 2]. The previously described variants are classified in benign, likely benign, variant of uncertain significant, likely pathogenic, and pathogenic. There are many databases that can be used.

The first question is the nomenclature. Considering that in the literature the variants may have different names, the nucleotide gene numbering may be different and multiples transcripts and variants can be identified in a different way. The genome build, gene name, reference transcript, cDNA, and amino acid level annotations that will be used must be previously defined in the laboratory [3]. For nomenclature, the Human Genome Variation Society guidelines can be followed [4].

Different databases provide information about the pathogenicity of known variants. One of the most recommended is The Human Gene Mutation Database where we found mutations for each gene sorted into groups depending on the nature of variation. It also includes a reference to the phenotypic effect and links to publications where these mutations are described. Also the NCBI home page can be used.

1. Open the NCBI home page.
2. Select "Nucleotide" of the scroll on the left.
3. Write the name of the gene in the box and enter.
4. Select the genomic sequence from the blue square.
5. Click graphic.
6. Now the entire gene appears and the variant can be found. Fill in the box and click find. When the variant has been previously described a red mark appears in the position; click the mark. The available information including the rs number of the variant will appear. This page is linked to ClinVar an NCBI initiative aiming to share clinical information of genetic variants.

3.5 Prediction of Pathogenicity

For not previously described variants or for variants with uncertain significant the pathogenicity degree must be assessed. In this sense, there are two specific questions, if the variant alters the gene function and if the functional change modifies the phenotype or it results in a disease.

Firstly, characterizing the phenotype of the disease is needed, by reviewing the literature looking for the penetrance, expressivity, and age of onset or prevalence, and determining the inheritance and the strength of associations. It also could be important to analyze the type of mutations that are disease causing. When a new mutation appears the functional impact of the change must be assessed.

- When the change occurs in an exon it will be possible to predict an effect on the protein. It is necessary to study on one hand the nature of the exchanged amino acids and on the other hand the position in the protein. Knowing the amino acid nature helps to predict its involvement in the protein structure. The location of the aminoacid in the protein can predict whether the protein may lose functionality because sometimes affecting active sites, binding domains of other proteins, phosphorylation sites, etc.
- An insertion or a deletion of a base will cause a change in the reading frame and this will cause a change in the protein, or will create a stop codon that generates a truncated protein.
- The situation is different if the change is in an intron. According to the distance at which change occurs, the consensus sequence

set (splicing) can be destroyed and the adjacent exon could be skipped resulting in a protein that has lost some number of amino acids. Or it may occur that a cryptic site is activated in an intron and the exon produced will be larger and possibly with a change in the reading frame.

Considering that a variant is more likely pathogenic if it occurs more than expected by chance in affected individuals, the population control studies [5] should also be evaluated, looking for the frequency of the mutated allele in control populations; however, the presence of a variant in large population studies does not imply that it is benign since pre-symptomatic, asymptomatic, or young individuals can be included in these studies [3].

Although the strongest functional evidence is provided through functional studies, there are bioinformatics tools to in silico predict how mutations affect the functionality of the protein and subsequently their pathogenic effect. The sequence conservation may indicate regions of functional importance.

SIFT (Sorting Intolerant From Tolerant) predicts deleterious or non-tolerated SNPs on the premise that some amino acids tend to be conserved in a protein family and any substitution at these positions would affect protein function and thus have a phenotypic effect. SIFT calculates the normalized probability in terms of SIFT score or tolerance index score for each mutation.

One of the most useful is PolyPhen that helps predicting the possible impact of replacing one amino acid in the structure and function of the protein. PolyPhen generates multiple sequence alignment of homologous protein structures for predicting if the substitution is probably damaging or benign. First the genome sequence must be translated to protein. Serial Cloner or a similar program can be used.

1. Open Serial Cloner.
2. Paste the genomic sequence in the box.
3. Select "Protein" from the tool bar.
4. Select "Paste translated DNA to new protein window."
5. The protein sequence will appear.
6. Change the base of the variant in the genomic sequence.
7. Generate the new protein.
8. Again click "protein" and "align protein."
9. Select both generated proteins and "local align."
10. Then click "highlight differences"; the amino acid changed will be remarked.
11. Open Polyphen-2.
12. Paste the protein sequence previously obtained from serial cloner, without the change, in the wild-type version.

Table 2
Amino acid code

1 Letter	3 Letter	Description
A	Ala	Alanine
C	Cys	Cysteine
D	Asp	Aspartic acid
E	Glu	Glutamine acid
F	Phe	Phenyalanine
G	Gly	Glycine
H	His	Histidine
I	Ile	Isoleucine
K	Lys	Lysine
L	Leu	Leucine
M	Met	Methionine
N	Asn	Asparagine
P	Pro	Proline
Q	Gln	Glutamine
R	Arg	Arginine
S	Ser	Serine
T	Thr	Threonine
V	Val	Valine
W	Trp	Tryptophan
Y	Tyr	Tyrosine

13. Introduce the position and the amino acid substitution (*see* Tables 1 and 2).
14. Fill in the query description.
15. Submit query.
16. The program will generate a report with the degree of pathogenicity foreseen for this substitution.

Other different platforms can be used to analyze mutation (*see* Table 3).

Finally, when the new variant has been validated and identified as a probable pathogenic, the family study should be analyzed to see how the variant segregates with the disease in the family (*see* **Note 11**).

Table 3
Platforms to analyze new genetic variants

Name	URL
NCBI	http://www.ncbi.nlm.nih.gov/
Exome variant server	http://evs.gs.washington.edu/EVS/
ENSEMBL	http://www.ensembl.org/index.html
Leiden open variation	http://www.lovd.nl/3.0/home
TCGA	http://projects.tcag.ca/variation/
Nucleotide NDBI	http://www.ncbi.nlm.nih.gov/nuccore/189027144?report=graph
Polyphen-2	http://genetics.bwh.harvard.edu/pph2/
SIFT	http://sift.jcvi.org/
Uniprot	http://www.uniprot.org/uniprot/
SWISS	http://bioinf.umbc.edu/dmdm/gene_prot_page.php?search_type=protein &id=83304912
PMUT	http://mmb.pcb.ub.es/PMut/
SSF	http://www.umd.be/searchSpliceSite.html
Maxent	http://genes.mit.edu/burgelab/maxent/Xmaxentscan_scoreseq.html
NNSPLICE	http://www.fruitfly.org/seq_tools/splice.html
HSF	http://www.umd.be/HSF/
Mutalizer	https://mutalyzer.nl/
1000 Genomes Project	http://browser.1000genomes.org
Gene reviews	http://www.ncbi.nlm.nih.gov/books/NBK1116/
Genesplicer	http://ccb.jhu.edu/software/genesplicer/
Mutation taster	http://www.mutationtaster.org/
Mutation asesor	http://mutationassessor.org/
Condel	http://bg.upf.edu/blog/tag/condel/
Align GVGD	http://agvgd.iarc.fr/agvgd_input.php/

It should be considered that variable expressivity, incomplete penetrance, and late age of onset could mask the genotype-phenotype correlation, and therefore unaffected family members should not contribute to the segregation information in these cases [3, 6].

In summary, to assess the clinical significance of a variant, all the available evidence should be carefully examined. Clinical data and functional evidence must be evaluated.

4 Notes

1. Before starting the validation procedure it is necessary to know the limitations of the reference methodology. For instance, certain copy number variations, VNTR, or bigger deletions cannot be confirmed by EC sequence.
2. Confirm that you have selected the appropriate version, since gene mapping differs in the different available versions.
3. "Exon in upper case" allows easily distinguishing between intronic and exonic sequences, as well as to identify the first ATG.
4. Depending on how the variant was previously identified, the position in the sequence can be found by the rs number or by the flanking sequences. In the first case you can directly use the same UCSC platform. If you have the flanking sequences you can paste your sequence in a Word page eliminating all paragraph marks, and then just search the flanking sequence. You can also use any available program for DNA sequence analysis.
5. In the event that more than one SNP are in the same gene, assess whether it can be included on the same amplicon. In this case the size range of amplicons must be considered. For long fragments, the last bases (further than 500 bp) could not be easily reached. In this case, to analyze the entire fragment, the amplicon could be sequenced in two different reactions by both primers. Amplicons shorter than 80 bp are not interpreted because the first bases of the fragment are unreadable.
6. PCR conditions can be standardized in the laboratory for routine assays. In this case, the *Tm* should be identical to allow simultaneous amplification of different sequences.
7. Be careful with the sense of the primer. When the primers are given in the 5′–3′ sense, the reverse primer needs to be turned around for being found in the sequence. For this you can use the Reverse Complement Program.
8. If the background noise impairs the correct interpretation, you should sequence the fragment again. Care must be taken with the PCR conditions, with the purification of the fragment and with the CE conditions.
9. Two different amplicons have been simultaneously amplified when two overlapping sequences are detected. Review the PCR or the purification conditions.
10. The letter N in the chromatogram indicates that the system is unable to assign a base. This could be due to noise or to a heterozygous site (Fig. 1). The heterozygous sites use two superposed peaks shorter than the flanking bases although this is not always true. You can repeat the sequencing of the fragment by using the opposite primer, to clarify it. Usually, the background

for a specific place, or sequence artefacts disappear when the amplicon is sequenced in the other sense.

11. In the family studies it is very important to exclude the possibilities of non-paternity and sample swap. In addition, a pathogenic association with a mutation can be due to a tight linkage with another pathogenic variant.

References

1. Richards CS, Bale S, Bellissimo DB, Das S, Grody WW, Hegde MR, Lyon E, Ward BE, Molecular Subcommittee of the ACMG Laboratory Quality Assurance Committee (2008) ACMG recommendations for standards for interpretation and reporting of sequence variations: revisions 2007. Genet Med 10:294–300
2. Kearney HM, Thorland EC, Brown KK, Quintero-Rivera F, South ST, Working Group of the American College of Medical Genetics Laboratory Quality Assurance Committee (2011) American College of Medical Genetics standards and guidelines for interpretation and reporting of postnatal constitutional copy number variants. Genet Med 13:680–685
3. Duzkale H, Shen J, McLaughlin H, Alfares A, Kelly MA, Pugh TJ, Funke BH, Rehm HL, Lebo MS (2013) A systematic approach to assessing the clinical significance of genetic variants. Clin Genet 84:453–463
4. Taschner PE, den Dunnen JT (2011) Describing structural changes by extending HGVS sequence variation nomenclature. Hum Mutat 32:507–511
5. Abecasis GR, Altshuler D, Auton A, Brooks LD, Durbin RM, Gibbs RA, Hurles ME, McVean GA, 1000 Genomes Project Consortium (2010) A map of human genome variation from population-scale sequencing. Nature 467:1061–1073
6. Caleshu C, Day S, Rehm HL, Baxter S (2010) Use and interpretation of genetic tests in cardiovascular genetics. Heart 96:1669–1675

Chapter 3

Introduction to the Gene Expression Analysis

Ignacio San Segundo-Val and Catalina S. Sanz-Lozano

Abstract

In 1941, Beadle and Tatum published experiments that would explain the basis of the central dogma of molecular biology, whereby the DNA through an intermediate molecule, called RNA, results proteins that perform the functions in cells. Currently, biomedical research attempts to explain the mechanisms by which develops a particular disease, for this reason, gene expression studies have proven to be a great resource. Strictly, the term "gene expression" comprises from the gene activation until the mature protein is located in its corresponding compartment to perform its function and contribute to the expression of the phenotype of cell.

The expression studies are directed to detect and quantify messenger RNA (mRNA) levels of a specific gene. The development of the RNA-based gene expression studies began with the Northern Blot by Alwine et al. in 1977. In 1969, Gall and Pardue and John et al. independently developed the in situ hybridization, but this technique was not employed to detect mRNA until 1986 by Coghlan. Today, many of the techniques for quantification of RNA are deprecated because other new techniques provide more information. Currently the most widely used techniques are qPCR, expression microarrays, and RNAseq for the transcriptome analysis. In this chapter, these techniques will be reviewed.

Key words Gene expression, qPCR, Microarrays, RNA sequencing, Transcriptome

1 Quantitative PCR

In 1984, Kary Mullis developed the Polymerase Chain Reaction (PCR). This technique amplifies a specific segment of DNA to obtain hundreds of millions of copies in few hours [1]. It was used for qualitative studies and it was not until 1992 when Higuchi et al. developed the quantitative Polymerase Chain Reaction (qPCR), which employs the PCR technique to gene expression studies.

To do this, they used the same material as for conventional PCR, a pair of specific oligonucleotides as primers; deoxynucleotide triphosphates (dNTPs); a reaction buffer; a thermo stable DNA polymerase; and they added a fluorochrome that fluoresces when excited [2]. Additionally, they designed a system capable of detecting in real-time PCR products accumulated. The system uses

María Isidoro-García (ed.), *Molecular Genetics of Asthma*, Methods in Molecular Biology, vol. 1434,
DOI 10.1007/978-1-4939-3652-6_3, © Springer Science+Business Media New York 2016

a camera that detects the increase in fluorescence that occurs when the ethidium bromide is intercalated in new strands of DNA formed in each cycle [3, 4]. Therefore, in the real-time PCR or qPCR, the processes of amplification and detection occur simultaneously in the same vial.

The PCR reaction consists in a series of cyclical temperature changes. Each cycle is divided into three stages [5]:

- Denaturation: separation of double-stranded DNA when subjected to 95 °C.
- Hybridization of primers: alignment of the primers to the DNA template at temperature around 50–60 °C.
- Elongation or Polymerization: binding of the corresponding dNTPs, to the DNA elongation chain at temperature around 68–72 °C.

Theoretically, if the reaction efficiency is 100%, the number of DNA molecules will double with each cycle. The reality is that efficiency in optimal conditions will be slightly lesser than 100% [6].

1.1 Methods of Detection

The qPCR technique allows the quantification of starting material (DNA, cDNA, or RNA), by using fluorophores. The fluorescence is measured in each cycle and is proportional to the amount of the PCR product.

On the one hand, fluorophores can be fluorescent dyes that nonspecifically bind to double-stranded DNA and produce a fluorescent amount that correlates with the DNA copy number. In this group we find dyes such as SYBR Green [7]. On the other hand, can be used fluorochromes attached to probes, which specifically hybridize to the amplified DNA strands. Thus, the reaction is more specific and the signal only is generated when the probe hybridizes with its complementary region. In this group, we find hydrolysis probes “TaqMan,” FRET probes, Beacon probes, and Scorpions probes [8].

1.1.1 Nonspecific Fluorescent Dyes

SYBR Green I is an organic compound with the chemical formula $C_{32}H_{37}N_4S$ [9]. It is associated with the DNA molecule by interacting with the minor groove. It is used in DNA staining for electrophoresis analysis of PCR products, or as a means of direct visualization of the PCR products in real time [9].

The melting curve analysis is necessary to detect problems of nonspecific binding of SYBR Green to DNA. Comparison of the melting curves allows an interpretation of the amplified products in PCR. Different products have different temperatures because the melting temperature depends on the size of the amplicon, the GC content, and the secondary and tertiary structures. At the end of the PCR reaction, a gradient from 50 to 95 °C is made to denature the double-stranded DNA. When the double-stranded DNA becomes single-stranded DNA, the decrease in fluorescence can be observed as peaks, represented by performing the second derivative of fluorescence. If the reaction is not specific, the PCR products show different melting curve peaks [10]. Since more than a decade there are other fluorescent dyes such as LC Green, ResoLight, EvaGreen, Chromofy SYTO, and BEBO [11].

1.1.2 Sequence Specific Probes

- TaqMan® probe: belongs to the group of hydrolysis probes and is based on the 5′-3′ exonucleasa activity of the Taq polymerase. The Taqman probe is a sequence complementary to a PCR product that is not part of the primers. The probe is labeled with a fluorophore that covalently bonds to its 5′ end called "donor" and with a quencher covalently bonded to its 3′ end also called "acceptor" whose function is to quench the fluorescence emitted by the fluorophore when is excited by the light. The 5′ exonuclease activity degrades the probe and releases the fluorophore that will emit fluorescence when not being close to the quencher. The liberated fluorophore is proportional to the amount of DNA amplified in the PCR. The probe should be close to a primer and the amplicon size should not be greater than 200 pair bases. Another drawback is that these probes do not allow melting curve analysis because the hydrolysis of the probe prevents its reutilization [12, 13].
- Molecular Beacons: they have a fluorophore covalently linked to 5′ called "donor" and a quencher covalently linked to 3′ called "acceptor"; they also have a hairpin-like secondary structure which brings the fluorophore to the quencher. When the probe specifically hybridizes to DNA, the distance between the fluorophore to the quencher is opened, allowing registering the signal of the annealing phase. The difficulty is in the probe designing because the secondary hairpin structure of the probe–amplicon hybrid should be more stable than that formed by the molecular beacon [12–14].
- FRET probe: two probes system in which one oligonucleotide contains a donor fluorophore and the acceptor fluorophore is in another oligonucleotide. When both probes bind, the fluorophores are close and the power is transmitted from the donor to acceptor emitting fluorescence. A good design of the two primers and the two probes is critical to obtain good results [12–14].
- Scorpion probe: it acts both as a probe and as a PCR primer and has hairpin structure. The molecule has a fluorophore and a quencher similar to molecular beacon probes that bind to amplicon by the same principle. The reaction leading to the fluorescent signal is immediate. This is because it is attached to primer and not collides with targeted region; therefore, signals are stronger [14].

- MGB probe (Minor Groove Binder): it is considered a variant of Taqman probe but it is not hydrolyzed and has the bases modified. The probe is labeled with a fluorophore in 3′ and in 5′ with a quencher. The Minor Groove Binder is a part of the probe that binds to minor groove of DNA. Minor Groove Binder allows protecting the probe from the 5′-3′ exonucleasa activity of the polymerase. When the probe hybridizes the fluorophore is separated of quencher and a fluorescent signal is generated. The fluorescent signal is directly proportional to DNA quantity. An advantage of these probes is that they are shorter and more stable. They include superbases which are bases modified to increase the temperature of melting curve [12, 13].

1.2 Quantitative PCR Characteristics

1.2.1 Advantages of the Technique

The main advantages of this technique are as follows [15]:

- Speed: the testing time is approximately 1 h.
- Simplicity: the assay requires a pair of primers, an enzyme, dNTPs, and optionally a probe.
- Convenience: does not require postamplification processing.
- Sensitivity: samples with very few copies of messenger RNA (mRNA) can be quantified.
- Specificity: a well-designed assay is specific for a single target.
- Robustness: a well-designed trial will give results in a wide range of reaction conditions.
- High performance: thousands of reactions can be carried out in a single experiment.
- Familiarity: the PCR is well known; their advantages and disadvantages are well understood.
- Cost: the price of reagents is affordable and with small reaction volumes, the costs decrease.

1.2.2 Basic Terms of qPCR

All systems similarly quantify and record fluorescent signal each cycle of the PCR. The reaction kinetics shows a representation forming a sigmoid shape. The graphical representation is defined by the background baseline, threshold, and threshold cycle [13] (*see* Fig. 1).

- Background: Fluorescence not specific of the reaction. The mathematical algorithm removes it.
- Baseline: noise level in early cycles of PCR where it is not detected an increase in fluorescence of PCR product. Determines basal fluorescence.
- Threshold: threshold value is set just above the baseline level where the exponential amplification begins. Threshold level can be determined automatically or manually.

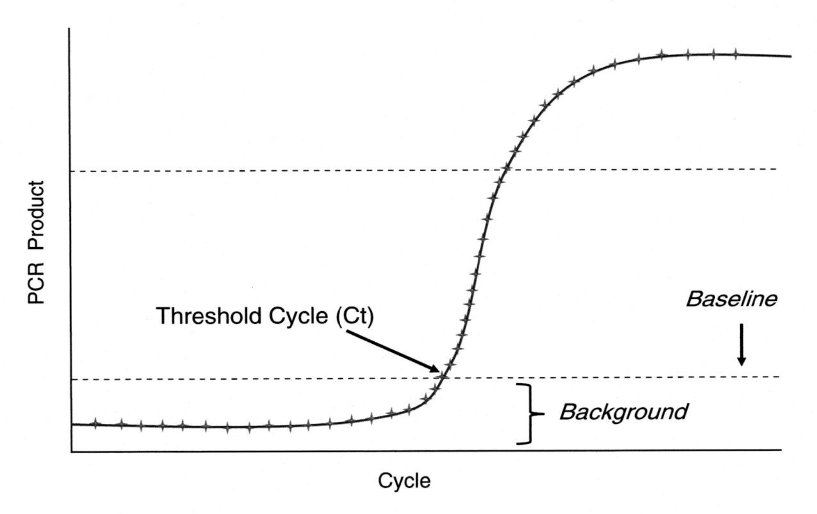

Fig. 1 Basic terms of qPCR

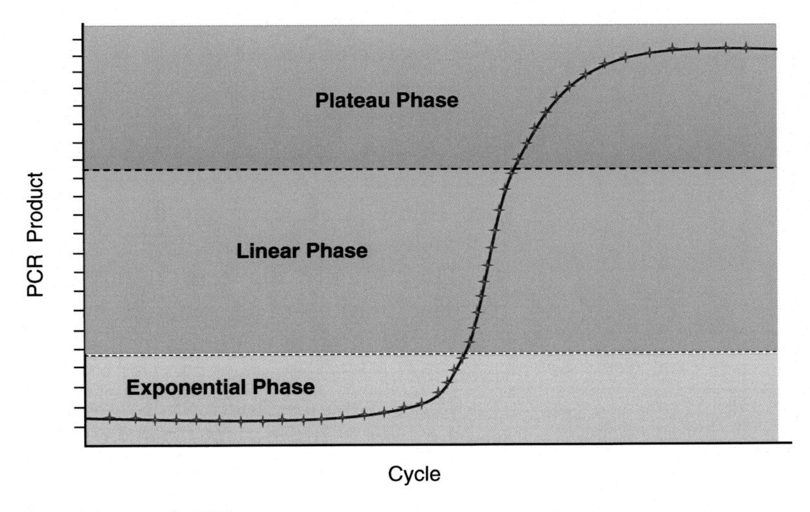

Fig. 2 Phases of qPCR

- Threshold cycle: the cycle at which the fluorescence exceeds the threshold. This cycle is used for relative quantification of gene expression.

1.2.3 Phases of qPCR

The qPCR reaction involves three phases that may be represented as a sigmoidal curve [16]. The three phases are as follows [12] (Fig. 2):

- Exponential phase: the amount of product is small, the PCR product is exponentially generated because of that the enzyme and reagents are not limited, so that reaction can reach the maximum efficiency. This exponential growth is difficult to detect because the quantity of product is insufficient. The amount of product at this stage is proportional to the amount of initial sample.

- Linear phase: the amount of product increases linearly because the quantity of enzyme and reagents begin to be limited so that reaction efficiency decreases.
- Plateau Phase: the reaction slows until the dNTPs and primers needed for new synthesis are depleted.

The cycle at which the fluorescence begins to exceed the background level is called the threshold cycle (Ct) and is the beginning of the logarithmic phase. Therefore, the Ct is inversely correlated with the amount of sample: the lower Ct, the greater cDNA amount.

1.2.4 Quantification

To quantify the expression of the gene under study there are two methods, called absolute quantification and relative quantification. In the first case, the absolute expression amount is expressed as number of copies obtained. In the second, the quantification is based on a calibrator, whose value is benchmark for all others, assigning it a value of 1 [17].

To perform absolute quantification is necessary to know the number of copies of target gene in a standard sample. Generally, this standard sample is a plasmid DNA or complementary DNA (cDNA) whose concentrations are measured by a spectrophotometer. In the assay, serial dilutions of the standard sample in which the target gene will be amplified should be prepared.

The computer registers the threshold cycles (Ct) for each sample. The dilutions of a standard sample show a standard curve that reflects the number of copies by interpolation of the Ct and allows quantifying the samples [12, 17].

There are three methods for relative quantification; the most widely used is the method 2-ΔΔCT [12]. This method assumes that the procedure doubles the DNA content in each cycle, that reaction efficiency is 100 %, and that a reference gene is expressed at a constant level between all samples [18]. This reference gene is a constitutive gene that is used as an endogenous control to correct intra and inter-assay variability.

The expression of the reference gene is changeless in the different samples of the assay as it is a gen which function is related to the cell maintenance, therefore, it is also called constitutive gene. The following formula expresses the ratio obtained from the relationship between the Ct values of the sample and the Ct values of the calibrator:

$$\text{Ratio} = 2^{-[\Delta \text{Ct sample} - \Delta \text{Ct calibrator}]}$$

$$\text{Ratio} = 2^{-\Delta\Delta \text{Ct}}$$

ΔCt sample is the difference between the Ct of the gene under study in sample minus the Ct of the reference gene in sample; ΔCt calibrator is difference between the Ct of the gene under study in the calibrator minus the Ct of the reference gene in the calibrator.

Another model for relative quantification was proposed by Pfaffl [19, 20]. In this model, different PCR efficiencies of the genes under study are taken into account, as shown in the following equation:

$$\text{Ratio}(\text{fold}) = \frac{(E\ \text{target gene})\ \Delta\text{Ct target}\ (\text{calibrator} - \text{sample})}{(E\ \text{reference gene})\ \Delta\text{Ct reference}\ (\text{calibrator} - \text{sample})}$$

In this model, it is necessary to know the efficiency of each pair of primers for each gene. The efficiencies are obtained from the slopes of standard curves obtained from serial dilutions according to the formula [16]:

$E = 10[-1/\text{slope}]-1$

The third model is called standard curve method or method-E that analyzes the efficiency of the gene under study and reference gene using standards, for which serial dilutions of a single sample are made. The standard curve method calculates the efficiency for each pair of primers of each gene [19, 21].

1.2.5 Selection of Reference Genes

Selection of the proper reference gene is a critical step to assess correctly data obtained by quantitative PCR. Most authors agree that the use of reference genes is the most effective and simple method for correcting bugs and glitches such as [8, 22]:

- Problems in the process of extraction, purification, or storage of RNA
- Bad performance of reverse transcription to cDNA synthesis
- Errors in pipetting or on transferring of material
- Polymerase inhibitors
- Poorly designed primers
- Inappropriate statistical analysis

Many qPCR experiments have been wrongly designed and are difficult to reproduce due to poor quality of data [8]. For this reason, in recent years the process of normalization of reference genes has become a recurring problem addressed by scientists. This has led to development of a variety of protocols and methodologies. Moreover, related publications showed big differences regarding the published information on development of their researches. This is being solved by guidelines for all publications associated with this methodology that are published in the MIQE guide (Minimum Information for Publication of Quantitative Digital PCR Experiments) [23].

Most publications agree that constitutive genes should show minimal variability among the tissues, cells, physiological conditions, or treatments under investigation [8, 24]. Therefore, it is necessary validate the reference genes for each tissue and treatment analyzed. Furthermore, it is desirable that the constitutive gene presents a threshold cycle (Ct) as close as possible to the problem gene, but it is not always possible, so that general recommendations should be followed. These recommendations do not advise choosing a constitutive gene with low expression (Ct > 30) or with high expression (Ct < 15) [8].

It is always recommended to use at least two reference genes, since the use of one can only result in relatively large errors [8, 23]. Although the use of a single reference gene is acceptable if it was previously tested in an experiment with similar conditions and was properly validated, it should be avoided to minimize the possible bias [8].

Several programs and models to select reference genes have been proposed with different statistical approaches and algorithms. Among the best-known applications are statistics to normalize, Normfinder, geNorm, and BestKeeper. Although there are approaches such as the ΔCt method or the classic ANOVA model for comparing the stability of the reference gene expression in the different study conditions by comparing the average Cts [25].

It also should be mentioned that many algorithms exist to study the variability of the constitutive gene expression, therefore the classification of the candidate reference gene might vary depending on the software or statistical technique used [26–29]. Finally, in selecting reference genes is advisable to select genes with different functions that do not have a common regulation that may affect its expression [8].

2 Microarrays

The Microarrays technique is based on complementarity between nucleic acid strands allowing detecting specific sequences by what is called hybridization [30]. The in situ hybridization was independently used for the first time by Gall and Pardue, and by John et al. in the same year. It was not used until 1986 to detect mRNA, by Coghlan [31]. In the late 1980s, Ekins et al. in University College London developed the first array for immunoassay studies, this array began to be manufactured and marketed in 1991 [32].

In the field of gene expression, it could study between 500 and 18,000 cDNAs. These human cDNAs were obtained from bacterial libraries and generally were radiolabeled with 33P-dNTPs. The most important limitations of macroarrays were the low density of probes onto the nylon support, large volumes of sample necessary for hybridization and that bacterial libraries could not be composed of pure colonies [33].

Technological efforts were focused on the miniaturization of arrays to overcome the limitations of the existing technology. In 1991, a group formed by Stephen Fodor, Leighton Read, Michael Pirrung, and Luberc Stryer fabricated a microarray. In 1993,

Stephen Fodor creates a spin-off that was dedicated to the microarrays development and in 1994 began the manufacture and sale of microarrays [34, 35].

Later, Patrick Brown used a glass support allowing a higher density of probes. Miniaturization decreases the amount of sample required for the study of gene expression, on the other hand, the use of radioactively labeled nucleotides was replaced by fluorophores [36]. In 1996, discloses the design, tools, and knowledge to let other research groups can make their microarrays in their laboratories. This information will boost the use of DNA microarrays [33].

The microarray, also known as DNA chip or biochip, is a solid support of glass, plastic, or nylon which is joined to oligonucleotides which sequences corresponds to all regions of the genome. The manufactures used the same technology used in the semiconductor chips but vertically placing million DNA strands on the support. Each oligonucleotide is placed in a specific area called "probe cell," where billions of copies of the oligonucleotide are found. These oligonucleotides are synthesized prior to bonding to the support or on the support. In the second case, there are many different methods: photolithography, phosphoramidite injection, or activation of precursors by an electric field [37, 38].

2.1 Types of Microarrays

Two types of microarrays are distinguished: expression microarrays, where specific RNA sequences are detected and genotyping microarrays for detecting specific DNA sequences [38]. By comparing the results of both arrays it could be established the relationship of polymorphisms with gene expression, which, among others, it could explain the different response to the treatment observed among patients with different genetic background.

2.1.1 Oligonucleotide Microarrays

Fragments of 25 specific base pairs are immobilized while chemically synthesized. By homogeneity, reproducibility, robustness, and high density are the most used since they can study up to 20,000 genes at once [38]. The main limitation is the cost of specific oligonucleotides selection and synthesis; it is more than three times more expensive than a cDNA microarray. The most important advantage is that photolithography can be used to direct synthesis of oligonucleotides which allows a high density of probes. Another advantage is that targeting sequence synthesis prevents cross-hybridization between sequences related genes. Furthermore, all the oligonucleotides of the microarray are of the same size, the same temperature melting, and the same concentration, therefore, the experimental variation is decreased, and the statistical power is increased [39].

2.1.2 cDNA Microarrays

The cDNA probes of 600–2000 base pairs are immobilized on a glass, nylon, or nitrocellulose base [38]. The main advantage is that these microarrays are cheaper than oligonucleotide microarrays. Another advantage is that it can be created from cDNA libraries, many of them of public domain.

The most important limitation is the data treatment due to the possibility of cross-hybridization that occurs between related genes. Another important limitation is that the variations of probe sizes and melting temperatures can diminish the statistical power [39].

On the other hand, these microarray experiments can be used in co-hybridization of two colors. These experiments allow direct comparison of mRNA abundance in two populations, although by this approach it is obtained comparative ratios rather than absolute levels of expression. On the other hand, the use of ratios reduces the inter-assay variation [39].

2.2 Methodology in Microarrays

The methodology of this technique includes the following steps [38, 40]:

- Extraction and preparation of RNA: the RNA is extracted from specific tissues, trying to obtain RNA with the highest purity and quality possible and thus avoiding a major source of variability.
- RNA amplification: the RNA is amplified by PCR to facilitate hybridization because the amount of mRNA in cells may not be sufficient.
- Reverse transcription: to convert mRNA into cDNA.
- Labeling the probes: the cDNA is fragmented and labeled with biotin. After, the fluorescent molecule that binds to biotin is added.
- Hybridization of the probes: the time required for complete hybridization is directly proportional to sample concentration. At the end of this process, the hybridized microarray is rinsed to remove unbound chains.
- Scanning of microarray: the fluorescent light detection indicates that hybridization has occurred at a specific point to a specific sequence. Reading is performed by a laser and the fluorescence is recorded by scanning. The fluorescence intensity is proportional to the amount of probe bound to each sample.

2.3 Limitations of Microarrays

The expression microarrays have produced a lot of information, but have limitations. Some of the main limitations of using microarrays are as follows [41, 42]:

- Limited dynamic range of detection: the detection of expression levels is limited to two or three orders of magnitude due to the background and the signal saturation. Thus, they are not suitable for studies of genes with low or very high expression.
- Reliance on existing knowledge: only can study known genome sequences, an errors in the database can worsen outcomes.

- Difficult comparison between experiments: they require complicated normalization methods for comparing expression levels from different experiments.
- Cross Hybridization: between genes of similar sequence; generates background and reduces the dynamic range.
- New transcripts can not be detected: new transcripts produced by alternative splicing can not be detected if the sequence is unknown.
- Necessity of validation: the microarrays need to be validated by qPCR for obtaining reliable microarray expression data. To do this, you must select and validate by using appropriate reference genes [43].

3 RNA Sequencing (RNAseq)

In 1964, Holley made the first complete sequencing of a gene. In the 1970s, Maxam and Gilbert developed a DNA sequencing technology based on chemical modification of DNA and subsequent cleavage at specific bases, while Sanger developed a DNA sequencing method based on the chain termination method. The Sanger sequencing was imposed by high efficiency and low radioactivity as the first-generation sequencing [44, 45].

The first automatic sequencer that used the Sanger method appeared in 1987 adopted the capillary electrophoresis as a more accurate and faster sequencing method. These sequencers have evolved from the 500 kilobases sequenced per day with the first model that appeared on the market until the 2.88 Megabases of the current model. These sequencers allowed to complete the human genome project in 2001 [45].

In 2005, the great revolution occurred in the field of DNA sequencing when the first sequencing of high performance appeared in the market. Several companies are now responsible for the development of high-throughput sequencers [45].

3.1 Transcriptome and RNA Sequencing

The high-throughput sequencers allow investigating the transcriptome. The transcriptome is the set of ribonucleic acids in the cell, including messenger ribonucleic acid (mRNA), transfer ribonucleic acid (tRNA), ribosomal ribonucleic acid (rRNA), small nuclear ribonucleicacid (snRNA), noncoding ribonucleic acids (ncRNA), and others. These RNAs are differently expressed according to the tissue, the physiological condition, or the stage of development [46].

The interpretation of the transcriptome complexity is a crucial objective to understand the functional elements of the genome and thus the functioning of the disease and its progression. In this sense, it has recently been shown that the amount of noncoding

DNA increases with the complexity of the organism, 0.25 % in the prokaryotic genome and 98.8 % in the human genome.

The existing level of complexity attached to the discoveries of endogenous small interfering RNA (siRNA), long interspersed noncoding RNA (lincRNA), transcription initiation RNA (tiRNA), microRNAs (miRNAs), transcription start site-associated RNA (TSSa-RNA) among others, represent parts of the transcription puzzle that we must decipher to understand how the genome works [45].

For this, the new RNA sequencing technology must start with cataloging all RNAs from mRNA; through the noncoding RNA to reach small RNAs determine transcription start sites and quantify changes in the expression levels of genes during development and in different conditions [41].

RNA sequencing is a technique in which sequenced fragments of 25–400 base pairs are used depending on the technology. To this end, a population of RNAs is fractionated and is transformed into a population of cDNA. These cDNAs are joined to adapters by one or both ends. Each cDNA molecule is sequenced to obtain short readings of one or both ends [41].

3.2 RNAseq Methodology

The design of a transcriptome study follows these steps [46]:

1. Selection of the tissue of interest in which the RNAs are to be studied.
2. Building the cDNA libraries.
3. Using a massive sequencing system.
4. Analyze the data using bioinformatics tools. Million short readings are obtained, which are mapped to a genome or transcriptome. The reads must be aligned to summarize all data. Finally, the data are normalized and the statistical tests are applied for studying differential gene expression, resulting in a classification of genes with their expression levels, the *p*-values, and the fold-changes [47].

There are several methods and programs to treat RNA sequencing data which aim not only estimating the differences in expression levels between samples, but also, detecting alternative splicing, analyzing the RNA editing, calculating the abundance of a transcript, etc. [47].

3.3 Advantages and Limitations of RNAseq

The RNAseq technology presents several advantages over other methodologies that aim to study gene expression:

- Resolution of a base: it can also detect changes in a single nucleotide (SNP) in the transcripts, microsatellites, isoforms, and allelic variants.
- Wide Dynamic Range: no upper limit for quantification and correlates with the number of sequences obtained, consequently

has a wide dynamic range of expression levels, estimated four to five orders of magnitude. So, it can analyze genes expressed at very low or very high levels that are not detected by other techniques. The low background-signal helps to improve the dynamic range.

- Short Readings: readings from 30 bp to allow accurate information about how to connect two exons.
- High accurate: it has been shown by validation studies with quantitative PCR.
- High levels of reproducibility
- Lower RNA sample: because there are no cloning steps.
- There is no need of reference genome: a reading can be performed without reference genome and the transcriptome can be de novo assembled. This is an advantage for species whose genome has not been sequenced yet [41, 45, 46].

However, this technique also presents some limitations as the high cost. It requires lots of resources compared to other techniques. The Data set is large and complex, the large amount of generated data makes the interpretation difficult, and usually a bioinformatic adviser is needed [46].

Acknowledgments

This work was supported by grants of the Junta de Castilla y León ref. GRS1047/A/14, GRS1189/A/15, and BIO/SA73/15; and by the project "Efecto del Ácido Retinóico en la enfermedad alérgica. Estudio transcripcional y su traslación a la clínica," PI13/00564, integrated into the "Plan Estatal de I+D+I 2013–2016" and cofunded by the "ISCIII-Subdirección General de Evaluación y Fomento de la investigación" and the European Regional Development Fund (FEDER).

References

1. Saiki RK, Scharf S, Faloona F et al (1985) Enzymatic amplification of beta-globin genomic sequences and restriction site analysis for diagnosis of sickle cell anemia. Science 230:1350–1354
2. Mallona I (2008) Selección de genes de normalización para RT-PCR cuantitativa en Petunia hybrida. (Normalization gene selection for quantitative RT-PCR in Petunia hybrida). Available via http://repositorio.bib.upct.es/dspace/handle/10317/723. Accessed 25 Nov 2014
3. Higuchi R, Dollinger G, Walsh PS et al (1992) Simultaneous amplification and detection of specific DNA sequences. Biotechnology 10:413–417
4. Higuchi R, Fockler C, Dollinger G et al (1993) Kinetic PCR analysis: real-time monitoring of DNA amplification reactions. Biotechnology 11:1026–1030
5. Clewley JP (1994) The polymerase chain reaction (PCR) for human viral diagnosis. CRC Press, Boca Ratón
6. Taylor S, Wakem M, Dijkman G et al (2010) A practical approach to RT-qPCR-Publishing data that conform to the MIQE guidelines. Methods 50:S1–S5

7. Huggett J, Bustin S (2011) Standardization and reporting for nucleic acid quantification. Accred Qual Assur 16:399–405
8. Kozera B, Rapacz M (2013) Reference genes in real-time PCR. J Appl Genet 54:391–406
9. Zipper H, Brunner H, Bernhagen J et al (2004) Investigations on DNA intercalation and surface binding by SYBR Green I, its structure determination and methodological implications. Nucleic Acids Res 32(12):e103
10. Sigma-Aldrich (2008) qPCR Technical Guide. Available via http://www.sigmaaldrich.com/content/dam/sigma-aldrich/docs/Sigma/General_Information/qpcr_technical_guide.pdf. Accessed 22 Nov 2014
11. http://www.gene-quantification.de/hrm-dyes.html
12. VanGuilder HD, Vrana KE, Freeman WM (2008) Twenty-five years of quantitative PCR for gene expression analysis. Biotechniques 44:619–626
13. Qiagen (2006) Critical factors for successful real time PCR. Integrated solutions-real time PCR applications. Available via http://jornades.uab.cat/workshopmrama/sites/jornades.uab.cat.workshopmrama/files/Critical_factors_successful_real_time_PCR.pdf. Accessed 28 Nov 2014
14. Leonard DGB (2007) Molecular pathology in clinical practice. Springer Science & Business Media, Berlin
15. Bustin SA, Kessler HH (2010) Amplification and detection methods. In: Kessler HH (ed) Molecular diagnostics of infectious diseases. De Gruyter, Berlín
16. Louw TM, Booth CS, Pienaar E et al (2011) Experimental validation of a fundamental model for PCR efficiency. Chem Eng Sci 66:1783–1789
17. Diez GO (2006) Técnicas de Genética Molecular II (Molecular Genetic Techniques II). In: Lasa A (ed) PCR cuantitativa (quantitative PCR). SEQC, Barcelona
18. Livak KJ, Schmittgen TD (2001) Analysis of relative gene expression data using real-time quantitative PCR and the 2(-Delta Delta C(T)) method. Methods 25:402–408
19. Pfaffl MW (2001) A new mathematical model for relative quantification in real-time RT-PCR. Nucleic Acids Res 29:2002–2007
20. Pfaffl MW (2004) Quantification strategies in real-time PCR. In: Bustin SA (ed) A-Z of quantitative PCR. International University Line, La Jolla
21. Bohla L, Dusanic D, Narat M et al (2012) Comparison of methods for relative quantification of gene expression using real-time PCR. Acta Agric Slov 100:97–106
22. Mallona I, Lischewski S, Weiss J et al (2010) Validation of reference genes for quantitative real-time PCR during leaf and flower development in Petunia hybrida. BMC Plant Biol 10:4
23. Bustin SA, Benes V, Garson JA et al (2009) The MIQE guidelines: minimum information for publication of quantitative real-time PCR experiments. Clin Chem 55:611–622
24. Valente V, Teixeira SA, Neder L et al (2009) Selection of suitable housekeeping genes for expression analysis in glioblastoma using quantitative RT-PCR. BMC Mol Biol 10:17
25. Podevin N, Krauss A, Henry I et al (2012) Selection and validation of reference genes for quantitative RT-PCR expression studies of the non-model crop Musa. Mol Breed 30:1237–1252
26. Fu W, Xie W, Zhang Z et al (2013) Exploring valid reference genes for quantitative real-time PCR analysis in *Plutella xylostella*. Int J Biol Sci 9:792–802
27. Gantasala NP, Papolu PK, Thakur PK et al (2013) Selection and validation of reference genes for quantitative gene expression studies by real-time PCR in eggplant (Solanum melongena L). BMC Res Notes 6:312
28. Paim RM, Pereira MH, Di Ponzio R et al (2012) Validation of reference genes for expression analysis in the salivary gland and the intestine of Rhodniusprolixus (Hemiptera, Reduviidae) under different experimental conditions by quantitative real-time PCR. BMC Res Notes 5:128
29. Tunbridge EM, Eastwood SL, Harrison PJ (2011) Changed relative to what? Housekeeping genes and normalization strategies in human brain gene expression studies. Biol Psychiatry 69:173–179
30. Southern E, Mir K, Schepinov M (1999) Molecular interactions on microarrays. Nat Genet 21:5–9
31. Eberwine JH, Valentino KL, Barchas JD (1994) In situ hybridization in neurobiology: advances in methodology. Oxford University Press, Oxford
32. McLachlan G, Do K, Ambroise C (2005) Analyzing microarray gene expression data. Wiley, Hoboken
33. Faiz A, Burgess JK (2012) How can microarrays unlock asthma? J Allergy 2012:241314
34. Affymetrix (2002) Affymetrix, Stanford University and incyte resolve patent oppositions and interferences. Available via http://investor.affymetrix.com/phoenix.

zhtml?c=116408&p=irol-newsArticle_pf&ID=362094. Accessed 21 Dec 2014
35. Times Higher Education (2006) Background memo on the winners of the European inventor of the year 2006 awards. Available via http://www.timeshighereducation.co.uk/news/background-memo-on-the-winners-of-the-european-inventor-of-the-year-2006-awards/203002.article. Accessed 22 Dec 2014
36. Shalon D, Smith SJ, Brown PO (1996) A DNA microarray system for analyzing complex DNA samples using two-color fluorescent probe hybridization. Genome Res 6:639–645
37. Lopez M, Mallorquín P, Vega M (2002) Microarrays y biochips de DNA, Informe de vigilancia tecnológica (DNA microarrays and biochips, technological surveillance report). Genoma España/CIBT-FGUAM
38. Daudén E (2007) Farmacogenética II. Métodos moleculares de estudio, bioinformática y aspectos éticos (Molecular study methods, bioinformatics and ethical aspects). Actas Dermosifiliogr 98:3–13
39. Alba R, Fei Z, Payton P et al (2004) ESTs, cDNA microarrays, and gene expression profiling: tools for dissecting plant physiology and development. Plant J 39:697–714
40. Lin SM, Johnson KF (2002) Methods of microarray data analysis II. Springer Science & Business Media, Berlin
41. Wang Z, Gerstein M, Snyder M (2009) RNA-Seq: a revolutionary tool for transcriptomics. Nat Rev Genet 10:57–63
42. Malone JH, Oliver B (2011) Microarray, deep sequencing and the true measure of the transcriptome. BMC Biol 9:34
43. Fernández AI, Óvilo C, Fernández A et al (2008) Luces y sombras del análisis de expresión génica utilizando microarrays. Un ejemplo en cerdo ibérico (Lights and shadows of gene expression analysis using microarrays. An example Iberian pig.) ITEA 104:99–105
44. Liu L, Li Y, Li S et al (2012) Comparison of next-generation sequencing systems. J Biomed Biotechnol 2012:251364
45. Costa V, Angelini C, De Feis I et al (2010) Uncovering the complexity of transcriptomes with RNA-Seq. J Biomed Biotechnol 2010: 853916
46. Santos CA, Blanck DV, de Freitas PD (2014) RNA-seq as a powerful tool for penaeid shrimp genetic progress. Front Genet 5:298
47. Oshlack A, Robinson MD, Young MD (2010) From RNA-seq reads to differential expression results. Genome Biol 11:220

Chapter 4

Real-Time PCR for Gene Expression Quantification in Asthma

Ignacio San Segundo-Val, Virginia García-Solaesa, and Asunción García-Sánchez

Abstract

The quantitative real-time PCR (qPCR) has become the reference technique for studying gene expression in recent years. The application of qPCR to the study of asthma provides very useful information regarding the gene expression mechanisms. The quantification of RNA from cDNA can be performed by using fluorescent dyes or specific sequence probes. Here, we describe the protocol to quantify gene expression levels using SYBR Green as fluorescent dye. The protocol starts with the RNA extraction, followed by reverse transcription to obtain cDNA, quantification and finally data analysis.

Key words Asthma, cDNA, Expression analysis, mRNA, qPCR, RT-qPCR (reverse transcription-qPCR)

1 Introduction

The quantitative PCR technique was developed by Higuchi et al. in 1992, as a modification of the polymerase chain reaction. They designed a system able to detect accumulated PCR products, as a fluorescence increase that reflected the quantity of DNA with ethidium bromide intercalated in it [1, 2]. Currently, a variety of probes and dyes are used to quantification.

To design a study of qPCR, initially, a careful selection of reference genes is needed. Another important aspect is the amount of RNA. This may limit the study of poorly expressed genes. In addition, RNA integrity is very important because the environmental RNases can degrade the mRNA [3, 4]. To quantify a gene it is necessary to transform mRNA into cDNA using a retrotranscriptase. We must guarantee a complete absence of contamination with genomic DNA before proceeding with the quantitative PCR. In this chapter a protocol of qPCR using SYBR Green is described.

María Isidoro-García (ed.), *Molecular Genetics of Asthma*, Methods in Molecular Biology, vol. 1434,
DOI 10.1007/978-1-4939-3652-6_4, © Springer Science+Business Media New York 2016

2 Materials

2.1 Storage of Blood Samples

1. Preservative RNA*later* (Ambion).

2.2 RNA Extraction

RiboPure™-Blood Kit (Ambion) containing:

1. Lysis solution.
2. Sodium acetate solution.
3. Acid-phenol:chloroform.
4. 100% Ethanol.
5. Wash solution 1.
6. Wash solution 2/3.

2.3 Treatment with DNase I

1. DNase I (Ambion).

2.4 Assessment of RNA Quantity and Quality

1. Micro-spectrophotometer.
2. *Agilent 2100 Bioanalyzer* (Agilent Technologies).
3. *Agilent* RNA 6000 Pico kit or *Agilent* RNA 6000 Nano kit.
4. *Agilent* RNA 6000 Pico Reagents or *Agilent* RNA 6000 Nano Reagents.
5. *Agilent* RNA 6000 Pico Ladder or *Agilent* RNA 6000 Nano Ladder.

2.5 Retro-transcription

SuperScript® III First-Strand Synthesis System for RT-PCR (Invitrogen):

1. Random hexamers (50 ng/μL).
2. dNTP mix (10 mM).
3. DEPC-treated water.
4. 10× RT buffer.
5. 25 mM NaCl2.
6. 0.1 M DTT.
7. RNase OUT™ (40 U/μL)/sample.
8. Superscript® III RT (200 U/μL)/sample.
9. RNase H.

2.6 Quantification

1. SYBR Green I Master (Roche Applied Science).
2. Reference gene and target gene primers (Table 1).
3. LightCycler 480 II (Roche Applied Science) or any real-time thermocycler.
4. PCR plates and corresponding optical sealing foils.

Table 1
Sequences of the primers used in this protocol

Accession number	Primer	Sequence
GAPDH P04406	*Primer* forward *GAPDH*	5′-CTCTGCTCCTCCTGTTCGAC-3′
	Primer reverse *GAPDH*	5′-ACGACCAAATCCGTTGACTC-3′
TBP P20226	*Primer* forward *TBP*	5′-GAACATCATGGATCAGAACAACA-3′
	Primer reverse *TBP*	5′-ATAGGGATTCCGGGAGTCAT-3′
PTGDR Q13258	*Primer* forward *PTGDR*	5′-GGCATGAGGCCTAAAAATGAG-3′
	Primer reverse *PTGDR*	5′-CCTTGACATCCTTAAATGCTCC-3′

3 Methods

Before starting the assay, define the characteristics of the experimental groups and calculate the sample size needed for the study.

3.1 Storage of Blood Samples

It is important to keep in mind that for RNA experiments it is recommended to work always with gloves RNase free without powder, and using a set of exclusive pipettes and filter tips to manipulate RNA samples.

In this protocol, the sample used is whole blood with anticoagulant EDTA; whole blood with citrate or heparin can also be used. It is important extracting the RNA immediately; if not, add a preservative to prevent that the lysis of blood cells releases RNases.

To store samples mix 450–500 μL of whole blood with 1300 μL of RNA*later*® and freeze in a 2 mL cryovial to −80 °C for a long-duration storage (*see* **Note 1**).

To immediately extract the RNA, add the lysis solution to the blood as indicated below.

3.2 RNA Extraction

1. Remove the RiboPure-Blood Kit from the refrigerator (at 4 °C) and preheat an aliquot of the elution solution to 75 °C.
2. Centrifuge the blood tube for 1 min in a microcentrifuge to maximum speed (*see* **Note 2**). Remove and discard the supernatant with a micropipette. The remnants of RNA*later*® can be eliminated gently inverting the vial on a filter paper.
3. Add 800 μL of lysis solution and 50 μL of the sodium acetate solution to the pellet.
4. Vortex for 1 min.
5. Add 500 μL of acid-phenol:chloroform (*see* **Note 3**).
6. Vortex for 1 min.

7. Leave the mixture at room temperature for 5 min.
8. Centrifuge for 1 min.
9. Separate the aqueous phase (supernatant) with a micropipette (*see* **Note 4**) and transfer to a 2 mL tube (*see* **Note 5**). Recover 1.2–1.3 mL approximately (*see* **Note 6**).
10. Add one-half volume of 100% ethanol to each tube (600 μL in this case).
11. Vortex the samples for 2–3 s.
12. Introduce one filter cartridge inside a 2 mL tube.
13. Add 700 μL of the sample to the filter cartridge.
14. Centrifuge for 10 s. Discard the flow-through from the tube and repeat until the entire sample is complete.
15. Add 700 μL of wash solution 1 to the filter cartridge and centrifuge for 10 s and eliminate the flow-through.
16. Add 700 μL of wash solution 2/3 to the filter cartridge and centrifuge for 10 s and eliminate the flow-through. Repeat this step (*see* **Note** 7).
17. Centrifuge for 1 min.
18. Add 30–50 μL of preheated elution solution to the center of the filter cartridge and close the cap. Leave the tubes at room temperature for 20 s, and then centrifuge for 30 s at maximum speed to recover the RNA (*see* **Note 8**).

3.3 Treatment with DNase I (See Note 9)

1. Apply 1/20th volume of DNase buffer to the sample.
2. Add 1 μL "DNase I" to the sample and mix pipetting up and down.
3. Incubate for 30 min at 37 °C in a thermocycler, a thermoblock, or a water bath.
4. Add a 1/20th volume of DNase inactivation reagent and incubate for 2 min at room temperature. Mix manually once or twice to resuspend the DNase inactivation reagent during the incubation (*see* **Note 10**).
5. Centrifuge in a microcentrifuge at maximum speed (RCF >16,000 ×*g*) for 1 min.
6. Recover the supernatant without taking the resin beads. These particles interfere with the assessment of the RNA quantity and quality.

3.4 Assessment of RNA Quantity and Quality

3.4.1 Micro-Spectrophotometer

Use 1–1.5 μL to RNA quantification. Purity is evaluated with two ratios: A260 nm/A280 nm and A260 nm/A230 nm. The ratio A260 nm/A280 nm must be greater than 1.8 and lesser than 2.0. Inorganic contaminants such as phenol or organic contaminants such as proteins can vary this ratio. The ratio A260 nm/A230 nm must be around 2.0. The residual phenol, the guanidine, or the carbohydrates alter this ratio [5].

3.4.2 Bioanalyzer

The measurement using the electropherogram Agilent 2100 Bioanalyzer provides the analysis of RNA integrity number (RIN). The RIN is an algorithm, which assigns a score from zero to ten (10 corresponding to the highest quality) based on the ribosomal RNA peaks and the extent of RNA degradation products [6]. The RNA quality should have a RIN > 7; a sample with RIN < 4 generates high variability in quantification cycle (Cq) [7]. To use an Agilent RNA 6000 Pico kit, 1 μL of a total RNA sample is needed ranging from 50 pg/μL to 250 pg/μL. For the Agilent RNA 6000 Nano Kit, 1 μL of a total RNA is used with a concentration ranging from 25 ng/μL to 500 ng/μL [8].

3.5 Retro-transcription

1. Thaw the reagents and maintain them in a cooler.
2. Vortex the random hexamers and the dNTP mix to homogenize.
3. In a 0.2 mL tube add 1 μL of random hexamers (50 ng/μL) per sample and 1 μL of dNTP mix (10 mM) per sample. Homogenize the pool and distribute in all tubes (*see* **Note 11**).
4. Add the DEPC-treated water and the RNA sample, until 8 μL in each tube. It is necessary to standardize the amount of RNA in all samples. For example RNA tubes are prepared to a final concentration of 500 ng.
5. If the amplicon is not designed between exons, you can check genomic DNA contamination by adding a control-RT without retrotranscriptase.
6. Incubate the PCR tube in thermocycler for 5 min at 65 °C to anneal the hexamers and leave it at 4 °C for 30 s minimum.
7. Prepare a pool with the following reagents (quantities per sample) (*see* **Note 11**):
   (a) 2 μL of 10× RT buffer.
   (b) 4 μL of 25 mM $NaCl_2$.
   (c) 2 μL of 0.1 M DTT.
   (d) 1 μL of RNase OUT™ (40 U/μL).
   (e) 1 μL of Superscript® III RT (200 U/μL).

   Briefly centrifuge to collect the components and mix by pipetting up and down before dispensing into the tubes that contain the sample, random hexamers, and dNTPs that are at 4 °C.
8. In a thermocycler incubate for 10 min at 25 °C, 50 min at 50 °C, 5 min at 85 °C, and 1 min at 4 °C.
9. Centrifuge to recover the liquid that remained in the top and add 1 μL of RNase H to each tube. Incubate for 20 min at 37 °C.
10. Continue immediately with the next step or store them at −20 °C (*see* **Note 12**).

3.6 Quantification

1. Design the primers, following these recommendations (*see* **Note 13**):
   (a) Amplicon length: 50–100 bp.
   (b) Primer length: 18–24 nucleotides.
   (c) GC content: 35–65 %.
   (d) Annealing temperature should be determined empirically and will depend on the melting temperature (*Tm*) of the primers. Design all primers to have approximately the same *Tm* (56–62 °C).
   (e) Specificity: Perform a BLAST to verify that they do not bind to more than one site.
   (f) SNP Database: Check that the sequences of the primers do not include SNPs.
2. As an example in this protocol we present the study of *PTGDR*, a gene related to asthma, by using two selected reference genes, *GAPDH* and *TBP*. The efficiency of the primers was previously assessed according to the protocol "validation of reference genes in expression analysis" (*see primers* in Table 1).
3. Turn on the computer of the system and enter the protocol to quantify with SYBR Green I (*see* Table 2).
4. From each sample calculate a triplicate with all components necessary for real-time amplification reaction, in a final volume of 15 μL:
   (a) 7.5 μL SYBR Green I Master 2X (*see* **Note 14**).
   (b) Water PCR grade.
   (c) 600 nM Primer forward.
   (d) 600 nM Primer reverse.
   (e) 25 ng/μL cDNA sample.

   The amounts and concentrations of the samples and primers must be adjusted based on intrinsic characteristics of genes under study and the research.

Table 2
Quantitative PCR protocol

Program	Temperature (°C)	Time	Ramp *T*ª	
Polymerase activation	95	10 min	4.4 °C/s	
Amplification				45 Cycles
Denaturalization	95	10 s	4.4 °C/s	
Annealing	60	10 s	2.2 °C/s	
Polymerization	72	10 s	4.4 °C/s	
Melting curve	97		0.11 °C/s	
Cooling	40	30 s	2.2 °C/s	

5. Prepare a pool for each gene with all components except samples (the minimum necessary for the assay is one reference gene, although at least two reference genes are recommended).
6. Vortex slightly and spin before pipetting.
7. Pipet the pools to each well of the plate, manually or with an automatic robot (*see* **Note 15**).
8. Add the samples and a calibrator sample for each gene. All wells should have the same concentration. Also include a no-template control (NTC) for each gene (*see* **Note 16**).
9. Seal the plate with the foil and give a spin in a centrifuge until RCF 450 × *g* (*see* **Note 17**).
10. Before quantification, use compressed air over the plate to eliminate powder that could interfere with the fluorescent signal.
11. Select the quantitative PCR protocol using SYBR Green I, and run the program.

3.7 Data Analysis

In the SYBR Green method it is recommended to calculate the *T*m (melting temperature) calling of the reactions, which gives the melting curve analysis. This analysis proves that only the specific PCR product has been amplified. In melting curve analysis the reaction mixture is slowly heated to 97 °C, causing the melting of the double-stranded DNA and a corresponding decrease of the SYBR Green I fluorescence.

The instrument monitors this fluorescence decrease and displays as melting peaks. Each melting peak represents the *T*m of a DNA product, which is determined by the length and the GC content of each fragment. If the PCR reaction generates only one amplicon the melting curve analysis will show only one peak. If primer-dimers or other nonspecific products are present there will be two or more melting peaks. The melting curve analysis is comparable with checking the length of a PCR product by gel electrophoresis.

1. For the Ct threshold, in the case that the standard deviation (sd) of the triplicates is >0.38, choose the two Ct data with a smaller ds by removing the outlier value [9].
2. Check that the NTCs have no Ct value.
3. Calculate the expression levels. The relative changes in the expression are analyzed by the $2^{-\Delta\Delta Ct}$ algorithm [10, 11]. For calculating ΔΔCt, you need the average of Ct values and corresponding sd.

   *Calculation of ΔCt value:

   $\Delta Ct = Ct\ target - Ct\ reference$

   *Calculation of ΔΔCt value:

   $\Delta\Delta Ct = \Delta Ct\ sample - \Delta Ct\ calibrator$

Differences in increments of a given gene expression calculated according to ΔΔCt method are typically expressed as a range (mean ± s.d.). The range to a target, relative to a calibrator sample, is calculated using the formula:

Relative expression = $2^{-\Delta\Delta Ct}$

In this algorithm we assume that efficiency is the same for all primers.

3.8 Check Genomic DNA (gDNA) Contamination

The gDNA contamination is checked by qPCR using primers that amplify genomic DNA of a reference gene.

1. Calculate triplicates for each sample in a final volume of 10 μL:
   - (a) 5 μL SYBR Green I Master 2×.
   - (b) Water PCR grade.
   - (c) 600 nM Primer forward.
   - (d) 600 nM Primer reverse.
   - (e) 25 ng/μL RNA sample.

     The amounts and concentrations of sample and primers must be adjusted based on intrinsic characteristics of genes under study.
2. Prepare a pool with all components except samples (*see* **Note 11**).
3. Pipette manually or with an automatic robot the pools to each well (*see* **Note 18**).
4. Add sample to each well, and mix by pipetting up and down. Add water to no-template control (NTC) (*see* **Note 16**) and cDNA to positive control (*see* **Note 19**).
5. Seal the plate with the foil and give it a spin in centrifuge until RCF 450 × *g* (*see* **Note 17**).
6. Use compressed air duster over the sealing foil to eliminate particles that interferes with the reading of fluorescent signal.
7. Run the quantitative PCR protocol for SYBR Green I (*see* Table 2).
8. Data analysis: To assess gDNA contamination, compare the average cycle threshold (Ct) (*see* **Note 20**) of the cDNA sample with average Ct of the RNA sample. The cDNA was amplified with expression primers and RNA with primers to amplify gDNA. The amplified cDNA provides a Ct; the RNA sample should not be amplified.

   Ideally, the difference between the cDNA and the RNA should be ten cycles (*see* **Note 21**).

4 Notes

1. Mix well the whole blood with RNA*later* to allow all cells to be in contact with the preservative. We recommend freezing the sample quickly to prevent the release of RNase by breaking the cellular membrane of blood cells.
2. All centrifugation steps in RiboPure-Blood extraction procedure should be done at an RCF of ~16,000 ×*g*, typically maximum speed on lab microcentrifuges.
3. Due to the toxicity and volatility of phenol-chloroform it is always necessary to work with gloves and in a chamber hood.
4. If the supernatant appears cloudy, nothing happens, it will not affect the quality of the extraction.
5. It is recommended to use safe-locked tubes to avoid spilling the sample by vortexing.
6. For less than 1 mL of the aqueous phase do not add twice phenol-chloroform (500 μL + 500 μL) because you can barely recover aqueous phase with a small quantity of RNA. You can add the lysis solution to the pellet (organic phase) again, and vortex to increase the lysis.
7. After the two washes and their corresponding centrifugations it is possible to stop the extraction protocol and start a second extraction. Meanwhile keep the filter cartridges at room temperature (20–25 °C) or preferably keep refrigerated at 4 °C. Then, centrifuge the filter cartridges from the two extractions for 1 min at maximum speed to eliminate residual alcohol before eluting RNA.
8. If you have to add the eluting solution to many tubes, do not allow the solution to cool.
9. The DNase treatment is necessary to eliminate gDNA when the amplicons are not designed in different exons. Use DNase I and an RT− control to check it.
10. It is important to mix well just before taking the resin beads (white particles) that quickly settle to the bottom of the tube.
11. It is recommended to perform a whole reaction mix according to the number of tubes plus an extra volume for the pipetting error (for example, each ten tubes add one extra volume). The pools save time and reactive.
12. For long term, store them at −80 °C to avoid degradation of the sample.
13. There are many documents with small differences in the design of primers; this is one example of many others.

14. It is recommended to follow this order when adding the reaction components. SYBR Green I is a fluorescent dye; it is recommended to protect it from light during the thaw. It is also important to keep separate the forward and reverse primers to avoid the possibility of cross-dimmers.
15. The cDNA samples should be kept on ice while pipetting; when you finish you can refreeze it. Room temperature favors the cDNA degradation increasing the Ct and providing greater variability.
16. The no-template control (NTC) carries PCR-grade water. Use a new vial of PCR-grade water in every plate. It is important to identify whether dimers are formed.
17. To seal the foil you can use a flexible reusable squeegee-type applicator.
18. If you take too long to pipetting, slightly vortex and spin to homogenize pool.
19. The RNA samples should have the same concentration per well than the positive control (cDNA).
20. Depending on the real-time instrument the cycle at which fluorescence from amplification exceeds the background fluorescence has been referred as Cp (crossing point), Ct (cycle threshold), or Cq (quantification cycle). Cq is recommended in the MIQE Rules [12].
21. Differences of ten cycles between cDNA and RNA indicate that the genomic DNA is about 0.1 % of amplified product. Treat the sample with DNase I again, when less than ten cycles of difference is detected.

Acknowledgments

This work was supported by grants of the Junta de Castilla y León ref. GRS1047/A/14, GRS1189/A/15, and BIO/SA73/15, and by the project “Efecto del Ácido Retinóico en la enfermedad alérgica. Estudio transcripcional y su traslación a la clínica,” PI13/00564, integrated into the “Plan Estatal de I+D+I 2013–2016” and cofunded by the “ISCIII-Subdirección General de Evaluación y Fomento de la investigación” and the European Regional Development Fund (FEDER).

References

1. Higuchi R, Dollinger G, Walsh PS et al (1992) Simultaneous amplification and detection of specific DNA sequences. Biotechnology 10:413–417
2. Higuchi R, Fockler C, Dollinger G et al (1993) KineticPCR analysis: real-timemonitoring of DNA amplification reactions. Biotechnology 11:1026–2030

3. Valente V, Teixeira SA, Neder L et al (2009) Selection of suitable housekeeping genes for expression analysis in glioblastoma using quantitative RT-PCR. BMC MolBio 10:17
4. Glare E, Divjak M, Bailey M et al (2002) ß-Actin and GAPDH housekeeping gene expression in asthmatic airways is variable and not suitable for normalizing mRNA levels. Thorax 57:765–770
5. Thermo Fisher Scientific (2013) http://www.nanodrop.com/Library/T042-NanoDrop-Spectrophotometers-Nucleic-Acid-Purity-Ratios.pdf Accessed 10 Apr 2015
6. Udvardi MK, Czechowski T, Scheible WR (2008) Eleven golden rules of quantitative RT-PCR. Plant Cell 20:1736–1737
7. Fleige S, Pfaffl MW (2006) RNA integrity and the effect on the real-time qRT-PCR performance. Mol Aspects Med 27: 126–139
8. Agilent Technologies. http://www.genomics.agilent.com/en/Bioanalyzer-DNA-RNA-Kits/RNA-Analysis-Kits/?cid=AG-PT-105&tabId=AG-PR-1172. Accessed 5 Apr 2015
9. Applied Biosystems. Guide to performing relative quantitation of gene expression using real-time quantitative PCR. http://www3.appliedbiosystems.com/cms/groups/mcb_support/documents/generaldocuments/cms_042380.pdf. Accessed 2008
10. Livak KJ, Schmittgen TD (2001) Analysis of relative gene expression data using real-time quantitative PCR and the 2–ΔΔCT method. Methods 25:402–408
11. Zhang JD, Biczok R, Ruschhaupt M (2011) The ΔΔCt algorithm for the analysis of quantitative real-time PCR (qRT-PCR). Bioconductor 2:11, http://bioconductor.org/packages/release/bioc/html/ddCt.html
12. Bustin SA, Benes V, Garson JA et al (2009) The MIQE guidelines: minimum information for publication of quantitative real-time PCR experiments. Clin Chem 55:611–622

Chapter 5

Validation of Reference Genes in mRNA Expression Analysis Applied to the Study of Asthma

Ignacio San Segundo-Val and Catalina S. Sanz-Lozano

Abstract

The quantitative Polymerase Chain Reaction is the most used technique for the study of gene expression. To correct putative experimental errors of this technique is necessary normalizing the expression results of the gene of interest with the obtained for reference genes. Here, we describe an example of the process to select reference genes. In this particular case, we select reference genes for expression studies in the peripheral blood mononuclear cells of asthmatic patients.

Key words Constitutive gene, Expression analysis, Reference gene, Normalization, qPCR, gene, Validation reference gene

1 Introduction

Most authors support that normalization of gene expression data with reference genes is the easiest method to control the experimental errors introduced at every stage of expression analysis. Reference gene must have a different sequence to the gene under study [1]. Normally, a proper normalization requires an internal control (IC), a reference, or reference gene. This reference gene is so named because it is assumed that their expression levels are maintained at a constant level in the different physiological conditions under study. In fact, these genes have functions that deal with the maintenance of the cell [2].

However, several studies mention that reference genes show some variation in their expression levels under some conditions, so they are not totally stable [2, 3]. For this reason, each researcher has to find candidate genes in each tissue and validate the genes that best adapts to the study conditions [4]. The reference gene can vary depending not only on the tissue and the treatment conditions, but also on the organism, disease, and many other factors [1]. In addition, normalize with a single reference gene may cause errors; therefore, it is always advisable to use at least two reference genes [5].

María Isidoro-García (ed.), *Molecular Genetics of Asthma*, Methods in Molecular Biology, vol. 1434,
DOI 10.1007/978-1-4939-3652-6_5,

The main problem to be considered in reference genes selection is the stability of the gene expression without using another reference, therefore, validation becomes a circulate problem [6].

There are many algorithms and procedures to select the best reference gene. The final election about which method to use should be made by the researcher whose target will be always get the more accurate and reproducible results.

2 Materials

2.1 RNA Extraction

1. MagNA Pure Compact Instrument (Roche Applied Science).
2. MagNA Pure Compact RNA Isolation Kit (Roche Applied Science):
3. Lysis Buffer.
4. Reagent Cartridges.
5. Tip trays.
6. DNase Solution.

2.2 Assessment of RNA Quantity and Quality

1. Micro-Spectophotometer.
2. *Agilent2100 Bioanalyzer* (Agilent Technologies).
3. *Agilent* RNA 6000 Pico kit or *Agilent* RNA 6000 Nano kit.
4. *Agilent* RNA 6000 Pico Reagents or *Agilent* RNA 6000 Nano Reagents.
5. *Agilent* RNA 6000 Pico Ladder or *Agilent* RNA 6000 Nano Ladder.

2.3 Check Genomic DNA Contamination

1. Gel electrophoresis device and a power supply to apply the electrical field.
2. Agarose Biotools Mb Agarose (Biotools).
3. TBE Buffer (Promega).
4. RedSafe™ of Staining Nucleic Acid Solution [20,000×] (iNtRON Biotechnology, INC).
5. DNA Loading Buffer (Lonza).
6. Gel Box for Gel Electrophoresis.
7. Molecular Weight Marker.
8. UV transilluminator VisiDoc-it™ Imaging System (Uplant).

2.4 Retro-transcription

SuperScript® III First-Strand Synthesis System for RT-PCR (Invitrogen, Life Technologies).

1. Random hexamers (50 ng/μL).
2. 10 mM dNTP mix.

3. DEPC-treated water.
4. 10× RT buffer.
5. 25 mM $NaCl_2$.
6. 0.1 M DTT.
7. RNase OUT™ (40 U/μL)/sample.
8. Superscript® III RT (200 U/μL)/sample.
9. RNase H.

2.5 Quantification

1. Thermocycler.
2. *The RealTime ready Human Reference Gene Panel* (Roche Applied Science) (Table 1).
3. 5× DNA Probes Master (Roche Applied Science).
4. Water PCR Grade (Roche Applied Science).
5. LightCycler 480 II or other real-time thermocycler (Roche Applied Science).
6. Plates and foils for thermocycler.
7. SYBR Green I Master (Roche Applied Science).
8. Reference genes primers.
9. Quantitative Polymerase Chain Reaction (qPCR) plate 96 well and foils (Roche Applied Science).

2.6 Selection of Reference Genes

The Real Time ready Human Reference Gene Panel includes the following reaction mixture per lyophilized well:

1. 200 nM of each primer.
2. 200 μM of dNTPs.
3. Taq polymerase 1 U.

3 Methods

To begin any expression study is necessary to define the sample selection criteria. A subsequent change in these criteria may invalidate the selected reference gene. Therefore it is always necessary to experimentally validate the reference genes that will be used in a study. It is not acceptable to use not internally validated reference genes except if the researcher shows evidence confirming that the expression of these genes does not change under the experimental conditions of the study [7].

3.1 RNA Extraction

RNA extraction from whole blood is immediately processed by the automated MagNA Pure Compact procedure for extraction of nucleic acids from a wide variety of samples (*see* **Note 1**). The equipment is based on a system of magnetic particles contained in

Table 1
Reference genes studied using *The RealTime ready Human Reference Gene Panel*

Abbreviation	Complete name
18S	Subunit 18S ribosomal
ACTB	Beta-actin
ALAS	Aminolevulinic acid synthase
β-2M	Beta-2-microglobulin
β-Globina	Beta-Globin
G6PDH	Glucose-6-phosphate dehydrogenase
GAPDH	Glyceraldehyde-3-phosphate dehydrogenase
GUSB	Glucuronidase beta
HPRT1	Hypoxanthine phosphoribosyltransferase 1
IPO8	Importin 8
PBGD	Porphobilinogen deaminase
PGK1	Phosphoglycerate kinase 1
PPIA	Peptidylprolyl isomerase A (cyclophilin A)
RPL13A	Ribosomal protein L13a
RPLPO	Ribosomal protein, 60s, P0
SDHA	Succinate dehydrogenase complex, subunit A
TBP	TATA-binding protein
TFRC	Transferrin receptor protein 1
YWHAZ	14-3-3 protein zeta/delta

a cartridge, which has a series of wells with the reagents needed for extraction. The RNA extraction system from the MagNA Pure Compact includes the following automated steps:

(a) Sample homogenate with Lysis buffer and proteinase K.

(b) Binding of nucleic acids to the surface of magnetic glass particles.

(c) Elimination of the cell debris by successive washes.

(d) Treatment of the sample with DNAse to remove the genomic DNA that remains in the same.

(e) Elution of RNA.

The protocol is as follows:

1. Turn on the MagNA Pure Compact system and select the RNA Blood protocol.
2. Reagent cartridges are introduced into the carousel after scanning their barcode, one reagent cartridge per sample (*see* **Note 2**).
3. The tip trays are placed in system, one tip tray per sample.
4. Add 20 μL of DNase into the IC site of the tube rack.
5. Add an aliquot of 200 μL of whole blood sample and enter the sample name.
6. Scan the barcode of the Elution tube and place it in the elution site of the tube rack. The RNA is eluted in 50 μL.
7. Run the program.

3.2 Assessment of RNA Quantity and Quality

The assessment of RNA quantity and quality can be done by using a microspectrophotometer and a bioanalyzer.

For microespectrophotometer, use 1–1.5 μL to RNA quantification to 260 nm. Purity is evaluated with 2 ratios; A260 nm/A280 nm and A260 nm/A230 nm. The ratio A260 nm/A280 nm must be between 1.8 and 2.0. Inorganic contaminants such as phenol or organic contaminants such as proteins can vary this ratio. The ratio A260 nm/A230 nm must be between 2.0 and 2.2, the residual phenol, guanidine, and carbohydrates alter this ratio [8].

The measurement by the electropherogram Agilent 2100 Bioanalyzer provides the analysis of the RNA Integrity Number (RIN). The RIN is an algorithm, which assigns a score from 0 to 10 (10 corresponding to the highest quality) based on the ribosomal RNA peaks and the extent of RNA degradation products. The RNA quality should have an RIN > 7, a sample with RIN < 4 generates high variability in Quantification cycle (Cq) [9]. For using Agilent RNA 6000 Pico kit, it is needed 1 μL of a total RNA sample ranging from 50 to 250 pg/μL. For the Agilent RNA 6000 Nano Kit, 1 μL of a total RNA with a concentration ranging from 25 to 500 ng/μL is used [10].

3.3 Check Genomic DNA Contamination

Genomic DNA contamination of the RNA samples can be checked by garose gel electrophoresis of RNA. You can also check the contamination in the retrotranscriptase step.

1. Prepare the electrophoresis gel using Biotools Mb Agarose at a concentration of 2% and 10× TBE Buffer diluted to 0.5× with distilled water.
2. Dissolve 2 g of agarose in 100 mL of 0.5× TBE by heating in a microwave.
3. Allow cooling to room temperature for 2–3 min.
4. Add 3 μL of RedSafeTM Staining Nucleic Acid Solution [20,000×] per 100 mL of gel. Add mixture to gel tray (*see* **Note 3**).

5. After 45 min the gel is solidified and is introduced into the gel electrophoresis device, filled with 0.5× TBE to proceed to load 5 μL of each sample with 1 μL of Loading Buffer (*see* **Note 4**).
6. Add the molecular weight marker.
7. Applied a current of 100 V for 20 min to the gel and analyzed the electrophoresis results on a UV transilluminator VisiDoc-it TM Imaging System. The putative genome DNA contaminations could be distinguished in the gel as high molecular size bands clearly different from the smaller ribosomal RNA bands.

3.4 Retro-transcription

1. Thaw the reagents and maintain the enzymes into a cooler.
2. Use Vortex to mix components.
3. In a 0.2 mL PCR tube, add 1 μL/sample of random hexamers (50 ng/μL) and 1 μL/sample of 10 mM dNTP mix. Prepare a master mix and dispense in all tubes.
4. To each identified tube, add DEPC-treated water and sample RNA to a final volume of 8 μL. It is necessary to standardize the amount of RNA in all samples that we want to convert into cDNA. For example, put 500 ng of RNA per tube (*see* **Note 5**).
5. Prepare a negative control tube without retotranscriptase (especially if the amplicon is not designed between exons). Without retrotranscriptase only a contamination of genomic DNA can be amplified.
6. Incubate PCR tube in thermocycler 5 min at 65 °C, followed by 30 s at 4 °C minimum.
7. Make a pool with: (*see* **Note 6**).
   - 2 μL/sample of 10× RT buffer.
   - 4 μL/sample of 25 mM $NaCl_2$.
   - 2 μL/sample of 0.1 M DTT.
   - 1 μL/sample of RNase OUT TM (40 U/μL).
   - 1 μL/sample of Superscript® III RT (200 U/μL).
   - Centrifuge briefly to collect components and mix by pipetting up and down before distributing.
8. In the thermocycler, incubate 10 min at 25 °C, 50 min at 50 °C, 5 min at 85 °C, and 1 min at 4 °C.
9. Centrifuge to recovery the liquid that remained in the top, add 1 μL of RNase H to each tube and incubate 20 min at 37 °C.
10. Continue or store the samples at −20 °C (*see* **Note 7**).

3.5 Quantification

As previously described, the genomic DNA contamination can be checked by making a quantitative PCR with RNA samples as negative controls and the cDNAs obtained from the retrotranscription as positive control. Any amplification detected in the qPCR using RNA as sample is an indicative of genomic DNA contamination.

1. Perform retrotranscription (explained in the previous section) to have the same cDNA concentration in all samples.
2. Prepare triplicates from each sample with all the components necessary for real-time amplification reaction. The final reaction volume is 10 μL:
   - 5 μL SYBR Green I Master 2×.
   - Water PCR Grade.
   - 600 nM Primer forward.
   - 600 nM Primer reverse.
   - 25 ng/μL Sample.

     The amounts and concentrations are different due to the intrinsic characteristics of each gene and should be empirically determined.
3. Make a pool with all components except samples, mix with vortex and give it a spin (*see* **Note 8**).
4. Pipette, manually or with an automatic robot, the pools to each well.
5. Add the sample to each well, mix by pipetting up and down. Add water to No Template Control (NTC) and cDNA to the positive control.
6. Seal the foil and spin the plate in a centrifuge at 450 ×*g* (*see* **Note 9**).
7. Before putting in the plate into the thermocycler, use a compressed air applier over the foil and the wells of the instrument to eliminate powder that may interfere with the reading of the fluorescent signal.
8. Run the Quantitative PCR Protocol for SYBR Green I (Table 2).
9. Data Analysis: To assess contamination, compare the average Cycle Threshold (Ct) (*see* **Note 10**) of the cDNA sample with the average Ct of the RNA sample and with the Ct of the NTC. Ideally, none amplification should be observed neither in the NTC nor in the RNA sample. In the practice, a Ct over 40 for the RNA amplification or a difference between the Cts of the RNA and cDNA samples over ten could be considered as no genomic DNA amplification in the samples. If amplification is detected in the RNA samples and the difference between the Cts of the RNA and cDNA samples is under ten cycles, it

Table 2
Quantitative PCR protocol for SYBR green I

Program	Temperature (°C)	Time	RampT[a]	
Polymerase activation	95	10 min	4.4 °C/s	
Amplification				45 Cycles
Denaturalization	95	10 s	4.4 °C/s	
Annealing	60	10 s	2.2 °C/s	
Polymerization	72	10 s	4.4 °C/s	
Melting curve	97		0.11 °C/s	
Cooling	40	30 s	2.2 °C/s	

is recommended an additional treatment with DNAse free of RNAse. Finally if amplification is detected in the NTC, the qPCR reagents should be examined for putative contaminations (*see* **Note 11**).

3.6 Selection of Reference Genes

To select an appropriate reference gene is necessary to study the expression variability in the studied conditions. To this end, it is recommended to analyze the stability of a set of reference genes whenever a new experiment is set up. The primers for the different reference genes should be designed or a predefined reference gene panel could be employed. Here, it is shown a protocol using a predefined panel.

Each sample is analyzed per triplicate using The Real Time ready Human Reference Gene Panel containing 19 reference genes, three positive controls, and two negative controls. This panel is designed to detect a concentration ranging from 50 pg/reaction to 50 ng/well. The 19 Reference genes analyzed are described in Table 1.

1. Make a pool with all components except samples (*see* **Note 12**). Add per sample:
   - 14 μL of water PCR Grade per well.
   - 4 μL of DNA Probes Master 5× LightCycler® per well
2. Vortex slightly and give it a spin before pipetting.
3. Pipette, manually or with an automatic robot, 18 μL of the pool to each well of the plate (*see* **Note 13**).
4. Add 2 μL of cDNA of a sample to all the wells of the 19 reference genes, positive controls, and negative control cDNA. It is important that all wells have the same concentration; 2 μL are added to the negative control RNA (*see* **Note 14**). The reaction mixture of each well has a final volume of 20 μL with 0.8 ng of cDNA (*see* **Note 15**).

Table 3
Quantitative PCR program for *The Real Time ready Human Reference Gene Panel*

Program	Temperature (°C)	Time	RampT[a]	
Polimerase activation	95	10 min	4.4 °C/sg	
Amplification				45 cycles
Denaturalization	95	10 sg	4.4 °C/sg	
Annealing	60	30 sg	2.2 °C/sg	
Polymerization	72	1 sg	4.4 °C/sg	
Cooling	40	30 sg	2.2 °C/sg	

5. Seal the foil and spin the plate in a centrifuge at 450 × *g* (*see* **Note 16**).
6. Before you put the plate into the thermocycler use a compressed air applier over foil and the instrument wells to eliminate any powder that may interfere with the reading of the fluorescent signal.
7. Run the quantitative PCR Program for The Real Time ready Human Reference Gene Panel (Table 3).

3.7 Data Analysis

Check that the NTCs not have Ct value.

Remove the triplicate Ct outlier (if any) by electing the two Ct with a standard deviation smaller than 0.38 [11].

Calculate the expression levels. The relative changes in the expression are analyzed using 2 –ΔΔCt algorithm [12, 13]. For calculating ΔΔCt, you need the average of Ct values and corresponding standard deviations.

* Calculation of ΔCt value:

ΔCt = Ct target – Ct reference

* Calculation of ΔΔCT value:

ΔΔCT = ΔCt sample – ΔCt calibrator

Differences in increments of a given gene expression calculated according ΔΔCT method are typically expressed as a range (mean ± standard deviation). The range to a target, relative to a calibrator sample is calculated using the formula:

$$\text{Relative expression} = 2^{-\Delta\Delta CT}$$

In this algorithm we assume that the efficiency is the same for all genes primers.

Use a statistical program to calculate the average Ct for each of the 19 reference genes of the panel together with its standard deviation (sd). These calculations should be made in the two conditions under analysis, for instance in the patients and in the controls group. To assess expression variability between the two conditions the coefficient of variation (cv) analysis can be used (*see* **Note 17**). Finally, select reference genes with less variability in the average Ct, not having differences between the studied groups and with the nearest cycle to the target gene.

3.8 Efficiency of Primers

To complete the validation, it is necessary to perform a standard curve to check the efficiency of all primers (*see* **Note 18**).

1. Take a cDNA sample and perform tenfold serial dilutions (*see* **Note 19**). A Minimum of five serial dilutions should be made (*see* **Note 20**).
2. Make triplicate reactions with all components for each sample in a final volume of 10 μL:
   - 5 μL SYBR Green I Master 2×.
   - Water PCR Grade.
   - 600 nM Primer forward.
   - 600 nM Primer reverse.
   - 25 ng/μL Sample.

   The amounts and concentrations are different based on intrinsic characteristics of the genes.
3. Make a Pool with all components except the cDNA samples, vortex and give it a spin (*see* **Note 8**).
4. Pipette, manually or with an automatic robot, the appropriate volume of the pools to each well.
5. Add cDNA dilutions to each well; mix by pipetting up and down. Do not forget to put a NTC with water.
6. Seal the foil and give a spin to the plate in a centrifuge at RCF 450 × *g* (*see* **Note 9**).
7. Before putting the plate in the thermocycler, use a compressed air applier over foil and the instrument wells to eliminate powder that may interfere with the reading of the fluorescent signal.
8. Run the quantitative PCR Protocol for SYBR Green I (Table 2).
9. Data Analysis: Calculate the average Ct of the triplicates of each dilution. After, perform a linear regression to obtain the equation of the line and the Pearson correlation coefficient (r) or the coefficient of determination (r^2). The r and r^2 coefficients indicate how the data fit the linear regression, it is recommended that the r coefficient is over 0.990 and the r^2 over 0.980 [14]. The equation of the line gives a slope to add to the formula for efficiency: $E = 10[-1/\text{slope}] - 1$

Table 4
Sequences of the primers used in this protocol

Accession number	Primer	Sequence
GAPDH P04406	*Primer* forward GAPDH	5′-CTCTGCTCCTCCTGTTCGAC-3′
	Primer reverse GAPDH	5′-ACGACCAAATCCGTTGACTC-3′
TBP P20226	*Primer* forward TBP	5′-GAACATCATGGATCAGAACAACA-3′
	Primer reverse TBP	5′-ATAGGGATTCCGGGAGTCAT-3′
PTGDR Q13258	*Primer* forward PTGDR	5′-GGCATGAGGCCTAAAAATGAG-3′
	Primer reverse PTGDR	5′-CCTTGACATCCTTAAATGCTCC-3′

Table 5
Primer efficiency data of genes related to asthma

Primer	Slope and intercept	Efficiency (%)	r^2 of calibration curve
GAPDH	Slope: −3.205 Intercept: 25.759	105	0.992
TBP	Slope: −3.459 Intercept: 33.525	94.5	0.999
PTGDR	Slope: −3.445 Intercept: 34.25	95.1	0.997

10. When the efficiency in each amplification cycle is 100% then the DNA quantity doubles. The reality is that the efficiency of the primers should be between 90 and 110% (*see* **Note 21**). The Sequences of the primers used in this protocol are shown in Table 4 and the data of primers efficiency of genes related to asthma in Table 5.

4 Notes

1. The automatic methods are suitable for large number of samples and for reference genes that are highly expressed in the cells. For poorly expressed genes manual extraction methods should be chosen.
2. Mixing the cartridge slightly to resuspend the magnetic particles, avoid bubbles in the remaining wells. If bubbles are formed give a light touch with your fingers to remove them.

3. Avoid bubbles to form, remove them with a comb.
4. If do not use it, store in refrigerator protected from light and dipped in 0.5× TBE.
5. If your Ct is still a little high, perform qPCR with more cDNA amount per well.
6. It is better to add the enzymes after all other components, and put the enzymes immediately back in the freezer.
7. If the samples are not frozen and remain at room temperature, the cDNA may be degraded and the Ct may increase having greater variability.
8. It is interesting to put a small remanent for 1 or ½ sample to not run out. The pools decrease the influence of pipetting errors in the results.
9. To seal the foil you can use a reusable flexible squeegee.
10. Depending on the system you use the data are shown as Cp (Crossing point), Ct (Cycle threshold), and Cq (Quantification cycle). In MIQE Guidelines the recommendation is Cq [7].
11. It is considered that the sufficient difference between cDNA and RNA amplifications is ten cycles because this indicates that there are 1000 times less genomic DNA than cDNA, so that the genomic DNA is about 0.1 % of amplified product.
12. Change the tip between samples to avoid polluting pool with lyophilized primers.
13. If it takes too long to pipette, vortex slightly and give a spin to homogenize pool.
14. The positive control informs about the degradation of initial RNA and the reverse transcription quality. The negative control detects the residual genomic DNA contamination. In this case, hydrolysis probes are used. If you use SYBR Green I, it is necessary to indicate the negative control Ct to validate the qPCR
15. The required amount is 50 pg/reaction to 50 ng/reaction.
16. To seal the foil you can use a reusable flexible squeegee type applicator.
17. The data can be statistically analyzed to detect whether there are significant expression differences between both conditions (in this case, the control group and group of patients diagnosed with asthma).
18. According to MIQE rules to validate the qPCR, not only check the efficiency of the primers, also know the limit of detection (LOD) and the Ct variation at LOD.
19. To make the standard curve use, for example, fourfolds or sixfolds serial dilutions. Consider that, the closer the points of the slope of the standard curve, the more variability you will have.

20. It is important to cover all possible sample concentrations that can be achieved during the study [14].
21. If efficiency is <90 % may be due to the inhibition of retotranscriptase, inhibition of Taq polymerase, a poor design of primers or to amplicons with secondary structures. If efficiency is >110 % may be due to primer dimmers or nonspecific amplicons [14].

Acknowledgments

This work was supported by grants of the Junta de Castilla y León ref. GRS1047/A/14, GRS1189/A/15, and BIO/SA73/15; and by the project "Efecto del Ácido Retinóico en la enfermedad alérgica. Estudio transcripcional y su traslación a la clínica," PI13/00564, integrated into the "Plan Estatal de I+D+I 2013–2016" and cofunded by the "ISCIII-Subdirección General de Evaluación y Fomento de la investigación" and the European Regional Development Fund (FEDER).

References

1. Kozera B, Rapacz M (2013) Reference genes in real-time PCR. J Appl Genet 54:391–406
2. Valente V, Teixeira SA, Neder L et al (2009) Selection of suitable reference genes for expression analysis in glioblastoma using quantitative RT-PCR. BMC Mol Biol 10:17
3. Podevin N, Krauss A, Henry I et al (2012) Selection and validation of reference genes for quantitative RT-PCR expression studies of the non-model crop *Musa*. Mol Breeding 30:1237–1252
4. Fleige S, Pfaffl MW (2006) RNA integrity and effect on the real-time qRT-PCR performance. Mol Aspects Med 27:126–139
5. Vandesompele JO, De Preter K, Pattyn F et al (2002) Accurate normalization of real-time quantitative RT-PCR data by geometric averaging of multiple internal control genes. Genome Biol 3:1–12
6. Nicot N, Hausman JF, Hoffmann L et al (2005) Reference gene selection for real-time RT-PCR normalization in potato during biotic and abiotic stress. J Exp Bot 421:2907–2914
7. Bustin SA, Benes V, Garson JA et al (2009) The MIQE guidelines: minimum information for publication of quantitative real-time PCR experiments. Clin Chem 55:611–622
8. Thermo Fisher Scientific (2013) http://www.nanodrop.com/Library/T042-NanoDrop-Spectrophotometers-Nucleic-Acid-Purity-Ratios.pdf. Accessed 10 Apr 2015
9. Udvardi MK, Czechowski T, Scheible WR (2008) Eleven golden rules of quantitative RT-PCR. Plant Cell 20:1736–1737
10. Agilent Technologies. http://www.genomics.agilent.com/en/Bioanalyzer-DNA-RNA-Kits/RNA-Analysis-Kits/?cid=AG-PT-105&tabId=AG-PR-1172
11. Applied Biosystems (2008) Guide to performing relative quantitation of gene expression using real-time quantitative PCR. http://www3.appliedbiosystems.com/cms/groups/mcb_support/documents/generaldocuments/cms_042380.pdf. Accessed 2008
12. Livak KJ, Schmittgen TD (2001) Analysis of relative gene expression data using real-time quantitative PCR and the 2–ΔΔCT method. Methods 25:402–408
13. Zhang JD, Biczok R, Ruschhaupt M (2011) The ddCt algorithm for the analysis of quantitative real-time PCR (qRT-PCR). Bioconductor 2:11, http://bioconductor.org/packages/release/bioc/html/ddCt.html
14. Taylor S, Wakem M, Dijkman G et al (2010) A practical approach to RT-qPCR-publishing data that conform to the MIQE guidelines. Methods 50:S1–S5

Chapter 6

Review of Methods to Study Gene Expression Regulation Applied to Asthma

Asunción García-Sánchez and Fernando Marqués-García

Abstract

Gene expression regulation is the cellular process that controls, increasing or decreasing, the expression of gene products (RNA or protein). A complex set of interactions between genes, RNA molecules, protein, and other components determined when and where specific genes are activated and the amount of protein or RNA produced. Here, we focus on several methods to study gene regulation applied to asthma and allergic research such as: Western Blot to identify and quantify proteins, electrophoretic mobility shift assay (EMSA) and chromatin immunoprecipitation (ChIP) to study protein interactions with nucleic acids, and RNA interference (RNAi) by which gene expression could be silenced.

Key words Allergy, Asthma, Chromatin immunoprecipitation (ChIP), DNA–protein, Electrophoretic mobility shift assay (EMSA), Gene regulation, siRNA, Western blot

1 Introduction

Gene expression is the process by which the genetic code (the nucleotide sequence of a gene) is used to direct protein synthesis. The flow of information from DNA to protein is tightly controlled by adjusting the amount and type of proteins produced. The amounts and types of messenger RNA (mRNA) molecules in a cell reflect its expression. Some genes are expressed continuously, as they produce proteins involved in basic metabolic functions; other genes are expressed occasionally, i.e., as part of the process of cell differentiation.

The process of gene expression involves two main stages, the transcription and the translation. In the transcription, the DNA information is copied into mRNA. The translation uses the mRNA to direct protein synthesis and a subsequent posttranslational processing of the protein exists.

Gene regulation is essential in all organisms and gives the cell the control over the structure and function and the cellular differentiation. It increases the adaptability of an organism, allowing the

María Isidoro-García (ed.), *Molecular Genetics of Asthma*, Methods in Molecular Biology, vol. 1434,
DOI 10.1007/978-1-4939-3652-6_6, © Springer Science+Business Media New York 2016

cell to express the needed protein in each moment. Several steps in gene expression process may be modulated, including the transcription, RNA splicing, translation, and post-translational modifications. The transcription regulation is the most extensively investigated mechanism [1, 2].

1.1 Transcriptional Regulation

The RNA polymerase is in charge of transcribing the DNA to mRNA. The promoter region of a gene attracts the enzyme RNA polymerase and begins its task of making an RNA copy of the gene. Several mechanisms must control how much mRNA must be produced. The regulatory DNA sequences are present in the promoter region located just upstream from the starting point for transcription (5′ end of the DNA) or located downstream of the mRNA (3′ end), more distant (enhancers) or even in the gene, oriented forward or reverse. These sequences play an important role in transcription by providing binding sites for regulatory proteins that affect RNA polymerase activity.

The transcription factors (TFs) are proteins that bind to these specific regulatory DNA sequences, to permit the control of the transcription from DNA to mRNA, by promoting (as an activator) or blocking (as a repressor) the recruitment of the RNA polymerase [1, 3–6]. Each TF has a specific DNA-binding domain that recognizes a 6- to 10-base pair motif in the DNA adjacent to the genes that they regulate [7, 8] and an effector domain. For an activating TF, the effector domain recruits RNA polymerase II to begin gene transcription.

The surface of the protein that recognizes a specific DNA sequence must fit tightly against the surface of the double helix in that region. Different proteins will recognize different nucleotide sequence. The protein–DNA interactions are highly specific and very strong, being considered among the tightest and most specific molecular interactions known in biology.

Frequently, DNA-binding proteins bind in pairs (homodimers) to the DNA helix. The dimerization doubles the area of contact with the DNA and increases the strength and specificity of the protein–DNA interaction.

Because two different proteins can pair in different combinations (heterodimerization), the dimerization also makes many different DNA sequences that can be recognized by a limited number of proteins. In eukaryotes, DNA is found in complex with proteins and RNA. This complex is the chromatin where the DNA is packaged around histone proteins forming the nucleosomes.

Each nucleosome consists in two copies of histones H2A, H2B, H3, and H4 (octamer) and 147 bp of DNA wrapped two times around the octamer [9]. The function of the chromatin is to package DNA into a smaller volume to fit in the cell, strengthen the DNA, allow mitosis and meiosis, prevents the DNA damage, and control gene expression and DNA replication. Histone proteins play an important role in the regulation of these processes.

Multiple residues found on histones are modified by processes such as acetylation and methylation, among others. These modifications interfere the contact between nucleosomes, "unravel" the chromatin and recruit nonhistone proteins [9, 10] and may up or down regulate the gene expression. These modifications that are inheritable and do not affect the DNA sequence are known as epigenetic regulation [11, 12].

The DNA methylation is a common method of gene silencing. DNA is methylated on cytosine nucleotides by methyltransferase enzymes in a CpG dinucleotide sequence called CpG islands. Abnormal methylation patterns are involved in human disease [13]. The histone acetylation mediated by histone acetyltransferase enzymes dissociates DNA from the histone complex, allowing the transcription. The DNA methylation and histone deacetylation are signals for DNA to be packed more densely, avoiding the transcription and decreasing the gene expression [14].

1.2 Post-transcriptional Regulation

The alternative splicing is the mechanism by which a single gene can code for multiple proteins. In this process, the primary transcripts synthesized by RNA polymerase contain sequences (introns) that will not be part of the mature RNA. These introns are removed before the mature mRNA leaves the nucleus.

The remaining region of the transcript includes the protein-coding sequences that are the exons, and they are spliced together to produce the mature mRNA. The proteins translated from alternatively spliced mRNAs will contain differences in their amino acid sequence and in their biological functions. The alternative splicing allows directing the synthesis of many more proteins than would be expected from its protein-coding genes [15].

All cells of a single organism carry exactly the same number of genes, so after the DNA is transcribed into mRNA, there must be a control of the mRNA that is translated into proteins. In eukaryotes, various post-transcriptional mechanisms add another control level over the complex systems that regulate gene expression. These mechanisms are the small inhibitory RNA (siRNA) and the microRNA (miRNA).

1.3 Post-translational Regulation

When a protein is synthesized, several changes can take place to determine whether the protein will be active. The proteins are synthesized by the ribosomes in which the mRNA is translated into polypeptide chains. These chains may be affected by post-translational modifications to form the mature protein. The modifications are covalent or enzymatic and can occur on the amino acid chains or at the C- or N-termini by introducing new functional groups such as phosphate, acetate, amide, or methyl groups [16].

The phosphorylation is the most common post-translational modification; the glycosylation consists in the addition of carbohydrate molecules that can promote protein folding and improve stability; and the lipidation is the attachment of lipid molecules of proteins to the cell membrane [17]. The Post-translational modifications of proteins can be detected by different techniques including Western blotting (WB) among others.

2 Methods of Study

2.1 Western Blotting

Western blotting (WB), also named protein blotting, Southwestern blotting, or immunoblotting, is the most common method to detect and quantify the protein expression. This method gives information about the size and identity of the protein of interest.

A cellular lysate formed by a mixture of proteins is separated through a sodium dodecyl sulfate (SDS) polyacrylamide gel using electrophoresis (SDS-PAGE) based on their molecular weight [18]. After the electrophoresis, it is transferred to an adsorbent membrane, which is incubated with labeled antibodies specific to the protein of interest. The unbound antibody is washed off leaving only the bound specific antibody, which is detected developing the film. Only a band of our specific protein should be detected. The amount of protein is proportional to the thickness of the band observed and can be easily calculated if a standard is used.

This method was first described in 1979 [18], and owes its name [19] to the geographic names of other techniques such as Southern [20] and Northern [21]. WB offers the following advantages: (a) the wet membranes are flexible and easy to handle, (b) the proteins immobilized on the membrane are readily and equally accessible to different ligands, (c) only a small amount of reagents are required for transfer analysis, (d) allows to screen different antibodies using one polyacrylamide gel, and (e) the immunoblot filter can be prepared for long time storage [22–24].

The protein blotting usefulness is based on its ability to provide a simultaneous resolution of multiple antigens in a sample, being a valuable method with many applications in research. Since its development, WB has evolved to solve different problems of the technique. One inherent problem associated with immunoblotting is the retention of high molecular weight protein. It was resolved with heat-mediated electrotransfer, consisting in transfer high molecular weight proteins entirely, rapidly, and without methanol [25]. An appropriate detection method is essential to the WB success. The most popular detection technique is the enzyme-conjugated antibodies used in conjunction with chemiluminescence [26, 27].

Western blot and subsequent immunodetection has been widely applied in the field of life sciences. A PubMed search using Booleans operators "AND" and the terms Western Blot and asthma, we found 1214 results on 3rd of February 2015. Only in 2014 more than 80 papers that have been used this method were published, which gives an idea of usefulness and value of this technique.

In the first studies in asthma, WB was used to characterize the wheat grain allergens involved in baker's asthma. In 1984, Shutton et al. performed a study in which analyzed the sera of 35 individuals with suspected allergy to the inhaled flour, and screened for the presence of immunoglobulin E (IgE) specific for wheat flour proteins. Sera from nine asthmatic bakers with high IgE levels were selected with the objective of purifying the allergen(s) involved in baker's asthma. The authors reported three purified wheat proteins identified as allergens for some but not all of the allergic bakers [28].

Several years after, Weiss et al. extracted proteins from wholemeal flour. The polypeptide composition (albumin/globulin, glialin, and glutenin) was analyzed by SDS-PAGE, transferred onto an immobilizing membrane and incubated with a pooled serum from asthmatic bakers. Bound IgE was demonstrated by autoradiography using 125I-labeled antihuman IgE. The study demonstrated that the serum of the bakers allergic to flour contained IgE antibodies bound to numerous polypeptides. The highest percentage of IgE binding was observed with a protein of 27 kDa albumin [29].

In 1992, Suphioglu C et al. studied the mechanism by which the rye-grass pollen causes asthma. The western blot analysis of isolated starch granules with specific monoclonal antibody and specific IgE from sera of grass-pollen allergic patients showed eight bands corresponding to major allergens. Authors conclude that starch granules released from rye-grass pollen provoke IgE-mediated responses causing asthma [30].

After these first works in which Western blot was used to identify and characterize possible allergens, this technique has become an indispensable tool, along with RT-qPCR, to analyze the expression of a particular protein in asthma. So we highlight the studies of Haitchi et al. with ADAM33 (A Disintegrin and Metalloprotease 33), which is strongly associated with asthma. Western blot analysis was undertaken to characterize ADAM33 protein expression in bronchial biopsies [31]. The presence of multiple isoforms of ADAM33, with different molecular weights, strongly suggested its importance in smooth muscle development and/or function, which could explain its genetic association with bronchial hyperresponsiveness [31].

Western blot is used too to characterize the state level of a protein. Signal transducer and activator of transcription (STAT) family proteins are dormant cytoplasmatic TFs, which become activated after phosphorylated, by kinases in response to binding of cytokine or growth factor receptors. The activated protein migrates into the nucleus and binds to specific promoter element to regulate gene expression [32, 33]. The IL-27-induced STAT1 phosphorylation was analyzed by WB in both healthy subjects and allergic asthmatic patients to investigate the mechanism by which CD4(+) T cells of asthmatic patients are resistant to IL-27-mediated inhibition. They conclude that differentiated Th2 cells can resist

IL-27-induced reprogramming toward Th1 cells by down regulating the phosphorylation of STAT1 [34]. IL-27-induced STAT1/3 phosphorylation has also been analyzed by Western blot in severe asthma [35].

Recently Mazzeo C. et al. investigated the secreted granules of the eosinophils involved in the inflammatory responses in patients with asthma. Western blot analysis were performed to demonstrate that some of these granules are exosomes and the discharge of exosomes to the extracellular media, increases after the IFN-γ stimulation, concluding that exosome production was augmented in asthmatic patients. Exosomes might play an important role in asthma and could be considered a biomarker of the pathology progression [36].

2.2 Protein: Nucleic Acid Interactions

The biological significance of protein interactions with nucleic acids is critical for many cellular functions, such as DNA replication, recombination, gene expression, cell differentiation, etc. The techniques most used to explore DNA–protein interactions are electrophoretic mobility shift assay (EMSA) and chromatin immunoprecipitation (ChIP) among others.

2.2.1 Electrophoretic Mobility Shift Assay (EMSA)

One important technique for studying gene regulation and the protein:nucleic acid interaction is EMSA. It is based in the observation that the electrophoretic mobility of a protein–nucleic acid complex in polyacrylamide gel is typically lesser than that of the free nucleic acid. Because the rate of DNA/RNA migration is retarded when bound to protein, the assay is also referred to as a gel shift or gel retardation assay. It is accepted that the technique was originally described by Garner and Revzin [37] and Fried and Crothers [38], although previous reports of this method can be found [39–42]. There are numerous variations of the technique according to each purpose [41].

The EMSA assays have the ability to resolve complexes of different stoichiometry or conformation. The source of the DNA/RNA-binding protein may be a whole cell or nuclear extract lysate, or a purified preparation. EMSA can be used qualitatively to identify sequence-specific DNA-binding proteins, such as TFs and quantitatively to measure thermodynamic and kinetic parameters [37, 38, 43, 44]. In an EMSA experiment, the protein–DNA complexes are electrophoretically separated on a polyacrylamide gel.

The molecules and its combinations move at different speed through the gel depending on the size and charge. The control lane (DNA probe without protein present) will contain a single band corresponding to the unbound DNA or RNA fragment. The lane with DNA and protein will contain another band that represent the complex of nucleic acid probe bound to protein which is "shifted" on the gel due to the larger and less mobile complex DNA/RNA–protein interaction.

Typically, short (20–50 bp) oligonucleotides bearing the binding sequence(s) of interest are used as target probe in EMSA. Traditionally, DNA probes were radiolabeled by incorporating 32P dNTPs and the gel was exposed to X-ray film after the electrophoresis. But a popular alternative detection system is a probe labeled with dNTP modified with haptens (biotin and digoxigenin). These can be visualized via secondary detection reagents such as streptavidin or anti-DIG antibodies in systems with enzymatic substrates to those used for WB, once the gel with the separated protein–DNA/RNA molecules are transferred onto an appropriate membrane.

Often, nonspecific competitor unlabeled nucleic acid is used as a blocking agent in the binding reaction to minimize the binding of nonspecific proteins to the target DNA. The unlabeled nonspecific competitor must be added to the reaction before the labeled protein, allowing a short incubation time. On the contrary, the specific competitor is an important control used to verify the specificity of a band resulting from the protein bound to the labeled probe. This specific competitor has the same sequence as the labeled probe or contains the consensus-binding motif for the target protein. The unlabeled specific competitors must be added to the reaction before the labeled probe, and after that the nonspecific competitor is incubated with the protein.

The shifts caused by the DNA/RNA–protein complex are characterized by a relative change in the mobility but does not identify the bound protein in the complex. The inclusion of an antibody that recognizes the protein present in the protein–nucleic acid complex can create a larger complex (antibody:protein:nucleic acid complex), and this method is known as "supershift" assay.

Some of the advantages of the method are that can detect low abundance of DNA-binding proteins and the possibility to test binding site mutations using many probe configurations with the same lysate. One limitation is that protein–DNA interaction in vitro is difficult to quantitate, it is necessary to perform an antibody-supershift assay to identify the protein in the complex. Another limitation is that EMSA is an in vitro method not within the cellular context.

Although EMSA has contributed to unraveling the protein–nucleic acid interaction, in recent years the development of in vivo methods and the use of high-throughput approach to detect biologically relevant interactions have become the alternative to in vitro EMSA assays. The chromatin immunoprecipitation followed of high-throughput sequencing (ChIP-seq) among others, have displaced the EMSA assays, but it still remains as a valuable tool to confirm the detected interaction.

A PubMed search using Booleans operators "AND" and the terms EMSA and asthma, we found 134 results on 3rd of February 2015. One of first works using this technique was reported in 1997.

Mori A et al. studied the mechanism of action of glucocorticoids (GCs) in the treatment of chronic asthma, accompanied by eosinophilia mediated by interleukin-5 (IL-5). The GC suppressed the IL-5 synthesis of T-cell clones activated via T-cell receptor (TCR) or IL-2 receptor (IL-2R). EMSA suggested that AP-1 and NF-Kappa B were the targets of GC actions on TCR-stimulated T cells [45].

Two years later, KH et al. investigated naturally occurring mutations in 5-Lipoxygenase (5-LO), an enzyme of the leukotriene biosynthesis, that modify functional expression of 5′-LO. In the promoter region of the *5LO* gene, there were different alleles with several Sp1 and Egr-1 putative binding motifs. EMSA and supershift analysis with nuclear extracts demonstrated the functionality of these binding motifs. The reporter assays performed indicated that these alleles were less effective than the wild-type allele in initiation *5LO* gene expression [46]. Papi A et al. analyzed the effect of desloratadine and loratadine on the rhinovirus-induced ICAM-1 expression, mRNA up regulation, and promoter activation. EMSA assay confirmed that desloratadine clearly reduced rhinovirus induction of the NF-kB binding [47] in infected A549 pulmonary cells.

In 2004, Oguma et al. reported four functional genetic variants of the prostanoid DP receptor (PTGDR) gene, previously associated with asthma, and localized at promoter region (−197 T>C, −441C>T, −549 T>C). These genetic variants presented differential profiles of *PTGDR* expression and the EMSA assays confirmed that these differences in transcription efficiency were linked with differences in TF binding.

Supershift and competition assays revealed that C/EBPb and members of the Sp and GATA families are differentially bound at *PTGDR* promoter region containing the SNPs and that the different SNPs modify the binding [48]. In 2011, our group researched *PTGDR* in asthmatic patients. Genetic variants were identified in patients and controls. Two new polymorphism, −95G>T and −613C>T, were incorporated to EMSA assays confirming modifications in the transcription-binding affinity pattern [49].

The Orosomucoid 1-like 3 (*ORMDL3*) gene was strongly associated with the development of asthma in genetic association studies. Qiu R. et al. examined the transcriptional regulation of *ORMDL3* identifying a minimal promoter region; in silico assays determined several STAT6 putative binding sites. EMSA demonstrated that STAT6 bound to its binding site at promoter *ORMDL3* [50]. Zhuang et al. characterized the mouse *ormdl3* gene promoter region by different approaches, including EMSA. In vitro EMSA assays identified a protein–DNA complex and the complex was supershift when it was incubated with the CREB antibody, validating the CREB-binding site in the *ormdl3* promoter region [51].

Another report in 2013 investigated the proteins that regulated the Th2 cytokines and Th2 cell differentiation. Using a TF-binding database and EMSA, YY1 was discovered as a candidate binding protein. The specific binding was confirmed by antibody-supershift antibodies and oligonucleotide competition experiments [52].

2.2.2 Chromatin Immunoprecipitation

The ChIP is a technique that like EMSA studies the interaction between proteins and nucleic acids. The main difference between them is that EMSA is an in vitro method, whereas ChIP is performed in living cells. ChIP has become a popular technique for identifying genome regions associated with specific proteins within their native chromatin context. Proteins are captured at the sites of binding to DNA and the method can be used to monitor in vivo the transcriptional regulation through the TFs–DNA interactions or histone modification [53–58]. This technique was originally developed by Gilmour and Lis to investigate the association of RNA polymerase II with transcribed genes in Escherichia coli and Drosophila [59–61]. They employed UV irradiation for covalently cross-linking proteins to DNA, in living cells. Later, it was replaced by formaldehyde to obtain a reversible cross-linkage [62].

In a typical ChIP, living cells are treated with formaldehyde to reversibly fix the DNA–protein interactions occurring in the cell. Then cells are lysed and chromatin is isolated from the nuclei and is sonicated or digested by nucleasase, to obtain sheared chromatin, to a range of 200–1000 bp. After that, protein–DNA complexes are immunoprecipitated by specific antibodies conjugated to beads. Finally, the cross-linkage is reversed and the DNA is purified to be ready for the analysis. End-point PCR or real-time PCR are popular methods of analysis to identify the immunoprecipitated DNA sequences. The quantitative PCR (qPCR) allows quantifying the amount of target DNA sequence [63].

Those methods are possible when it is known the target gene of the protein because it is necessary the design of primers for the detection of sequences. The combination of ChIP methods with cloning and other technologies as microarray (ChIP-chip) and ultra-sequencing (ChIP-seq) has allowed identifying unknown sequences in the genome-wide analysis of DNA–protein interactions [55–57].

The ChIP technique has evolved from its origins to the most advanced ChIP-seq in order to dilucidate global protein–DNA interactions. The most popular ChIP variation is ChIP-seq where DNA immunoprecipitated is subjected to massively parallel sequencing allowing simultaneously analyze millions of DNA fragments with efficiency and relatively low cost [64–66].

A PubMed search using Booleans operators "AND" and the terms chromatin immunoprecipitation and asthma revealed 65 results on 3rd of February 2015.

Many human promoter regions are known to exhibit genetic variability. These DNA polymorphisms regulate gene expression and have a considerable importance in asthma studies. To interrogate putative regulatory polymorphisms in vivo it is necessary to consider the native chromatin context and the natural haplotype combinations.

Knight JC et al. developed the haplo-ChIP, a method that combines ChIP with mass spectrometry to identify the in vivo protein–DNA binding associated with allelic variants of a gene. The method uses a genetic marker to differentiate between the alleles, the protein binding in the cell. The cross-linked chromatin fragments are subjected to PCR and primer extension to detect sequences containing a genetic marker. With the combination with mass spectrometry this approach allows a high-throughput genotyping. The application of this approach to the *TNF/LTA* locus identified functionally important haplotypes that correlate with allele-specific transcription of *LTA* [67, 68].

ChIP is typically employed to in vivo monitor the transcriptional regulation, through TF–DNA binding interactions. Most of studies combine EMSA assays with ChIP technology and reporter gene functional studies [48, 50–52, 67, 69]. In 2007, Robertson G et al. reported the first ChIP-seq to identify mammalian DNA sequences bound to TFs in vivo. They used ChIP-seq for mapping the STAT1 target in interferon-gamma (IFN-gamma)-stimulated and unstimulated HeLa S3 cells. 41,532 and 11,004 putative STAT1-binding regions were identified in stimulated or unstimulated cells, respectively. The ChIP-seq was enriched in sequences similar to known STAT1-binding motifs. Comparing with ChIP-PCR, the data sets suggested that ChIP-seq was more sensitive [66].

The first ChIP-seq in asthma was conducted in 2010 by Malhortra D et al. to study the nuclear factor E2 p45-related factor 2 (Nrf2) that binds to the cis-regulatory, antioxidant response element (ARE), and transcriptionally regulates the gene responses to the environmental stress. Authors described a genome-scale analysis of the regulatory network ruled by the NRF2 TF combining ChIP-seq and microarray-based gene expression profiling. The results confirmed the role of NRF2 in regulating the expression of protective genes that attenuate cytotoxicity in response to chemical toxins and in the direct regulation of cellular proliferation [70].

Recently Sharma et al. developed a GCs ChIP-seq experiment on a murine model to investigate whether GC genes are important during lung development and their role in asthma susceptibility and treatment response. Using gene expression profile of human and murine lungs, they identified a set of 232 genes of developing lungs, which were used to test the enrichment immunoprecipitated DNA from asthmatic subjects and controls before and after GC treatment [71].

Luca F et al. explored the mechanism of action of GC using different functional genomics approaches. GCs act mainly by activating the GC receptor (GR), which interacts with specific TFs. They performed ChIP-seq for GR and NF-kB in two lymphoblastoid cell lines treated with GC or vehicle control. As a result, the features of GR-binding sites differed for up- and downregulated genes. Analyzing these results with previous data of functional SNPs, the authors conclude that genetic variation at GR and interacting TF-binding sites influence in gene expression [72].

ChIP has been a powerful tool in the studies of histone modification in cells. In 2002, Valapour et al. employed the ChIP technology to define the role of histone acetylation in the regulation of *IL4* gene expression in human T lymphocytes. Deregulation of *IL4* has been related with allergic diseases and the alteration of chromatin structure is thought to play a role in regulating cytokine gene expression in Th2 cluster. ChIP assays showed that nucleosomes in the proximal *IL4* promoter are acetylated on T-cell activation, regulating the *IL4* gene expression [73]. In 2013, Cui ZL et al. explored the histone modifications of *NOTCH1* promoter in normal and asthmatic CD4+T cells. The analysis showed that the acetylation and trimethylation levels of histones of *NOTCH1* gene promoter were increased in asthmatic lung CD4+T compared with the control group. After treatment with Garcinol (GAR), a potent inhibitor of histone acetyltransferases, the histone acetylation in *NOTCH1* promoter decreased significantly and NOTCH1 expression was reduced. The effect of GAR also reduced significantly the levels of IL-4, IL-5, IL-13, and slightly the IFN-g level. Authors suggest that asthma is associated with changes in the epigenetic status of *NOTCH1* promoter, including the abnormal histone acetylation and methylation and they point to a possible application of GAR in the treatment of asthma [74].

2.3 RNA Interference

The RNA interference (RNAi) plays an important role in regulation of gene expression. It is a mechanism used by many organisms to silence the expression of genes that control different processes in the cell [75]. Certain (ds) RNA molecules inhibit gene expression of targeted genes with high specificity and selectivity. Different types of small ribonucleic acid molecules, microRNA (miRNA), small interfering RNA (siRNA), short hairping RNA (shRNA), and piwi RNA (piRNA) are involved in the RNAi. RNAi is a relevant research tool in cell cultures and in vivo experiments, because synthetic RNA introduced into cells can selectively silence specific target genes. RNAi was discovered in the nematode *Caenorhabditis elegans* and the work of Fire and Mello in 1998 was awarded with the 2006 Nobel Prize for physiology or medicine [76].

RNAi is activated when RNA molecules are present as double-stranded pairs in the cell. Double-stranded RNA (dsRNA) activates the RNAi pathway initiated by the enzyme Dicer [75], which cleaves long dsRNA molecules into short double-stranded fragments of ~20 nucleotide RNAs and 3′-ends overhangs named siRNAs (short interfering RNA) [77]. The siRNA then associates with RISC (RNA-induced silencing complex), a large protein complex (160 kDa) with a catalytic component the Argonauta proteins [78]. Inside the RISC, the siRNA is unwound into two single-stranded RNAs (ssRNAs), the passenger (sense) strand and the guide (antisense) strand. The passenger strand is degraded by cellular nucleases and the guide strand directs RISC to the complementary target mRNA sequence. This mRNA target is cleaved by Argonauta [79].

RNAi is a relevant mechanism for the regulation of gene expression present in plants and animals avoiding viral infections by knocking down viral mRNAs [80–82] and keeps jumping genes under control [83]. It is widely used both in basic and applied research as a method to study the function of genes and to develop promising gene therapies [84, 85].

Experimental RNAi in the mammalian system involves the introduction of siRNA, typically 21–23 bp duplexes, or shRNA molecules, that consist of a single strand having the sequence of the two desired siRNA strands connected by a no relevant sequence. The two complementary portions anneal intramolecularly allowing the RNA molecule to fold back on itself, creating a dsRNA molecule with a hairping loop. The Dicer enzyme cleaves the RNA structure into the desired siRNA molecule.

The siRNA can be prepared for several methods, such as chemical synthesis (the same chemistry used to generate DNA primers for PCR), in vitro transcription, siRNA expression vectors (containing siRNA or shRNA), and PCR expression cassettes. siRNA and shRNA are intrinsically different molecules and presents differences in the molecular mechanism of action and applications [86].

In a PubMed search using Booleans operators "AND" and the terms siRNA and asthma we found 333 results on 9th of February 2015. One of first works that used siRNA in asthma was reported by Li et al. in 2002. They investigated whether the regulation of prostaglandin D2 is mediated by RasGRP4 gene. A siRNA approach was used to evaluate the consequences of decreased expression of RasGRP4 in a rat cell line. The coding sequence of the target gene was scanned to identify a specific 21-nucleotide sequence downstream of an "AA" sequence with a 55 % GC content. After checking the specificity by BLAST searching, the RasGRP4 specific oligonucleotides sense and antisense were designed and the resulting siRNA duplex was introduced into an adherent cell line by lipotransfection. The PGD2 levels in control and siRNA treated cells were checked by Western blot showing a significantly decrease of PGD2 synthase when RasGRP4 is silenced. This data confirmed that RasGRP4 is required for the efficient expression of PGD2 in mast cells [87]. Since then numerous articles have been used this method for knocked-down genes and thus to analyze the consequences of their silencing [88–94].

Several articles report about the gene regulation controlled by miRNAs in asthma and allergic diseases. It is well known that miRNA can control the post-transcriptional regulation. It is postulated that an alteration in the miRNA expression could contribute to human respiratory diseases [95].

The mechanism of the miRNAs is the same that has been mentioned above. Changes in the expression of several miRNAs have been shown associated with development and/or improvement in animal models of asthma [96]. Williams et al. analyzed whether miRNA expression was differentially expressed in mild asthma and the effect of corticosteroid treatment. They found a miRNA expression profile specific to individual cell types but they could not find a link between miRNA expression profiles and asthma [97]. Jardim MJ et al. identified 66 miRNAs differentially expressed between patients with asthma and healthy donors [98]. Recently, Haj-Salem I et al. identified the miRNA-19a as an enhancer of cell proliferation in severe phenotype of asthma, suggesting that down regulate miRNA-19a expression could be explored as a potential new therapy to modulate epithelium repair in asthma [95].

2.3.1 Antisense and RNAi-Based Strategies in Asthma and Allergy

RNAi has become a standard method for in vitro knockdown of any target gene of interest. For the last two decades, scientists have been tried to translate the RNAi technology into clinical applications.

The main obstacle is the delivery method by which the siRNA enters to the target cell across the biological barriers. The nucleic acid is sensible to nuclease degradation and is not able to cross the biological membranes. Among all types of nucleic acids involved in RNAi: siRNA, long dsRNA, and shRNA, the siRNA is the most popular candidate. Because of its short size (typically 21–13 bp) it can avoid the interferon response in mammalian cells [99].

In 2004 was reported the first human clinical trial of RNAi therapy for the treatment of age-related macular degeneration with siRNA targeting VEGF-receptor 1, delivered intravitreally. Since then a number of clinical trials have been conducted in cancer and respiratory infection [99]. RNAi have used to investigate the possibility of treatment or prevent diseases affecting airways, such as lung cancer, respiratory diseases, airway inflammatory diseases, and fibrosis cystic [99]. In the lungs, siRNA delivery could be carried out by local delivery or systemic delivery, being preferable the local delivery in order to avoid systemic side effects. Local delivery can be achieved by inhalation, intranasal, and intratracheal route. Inhalation and intranasal route are both the preferred systems because the siRNA formulation can be aerosol or dry powder. However, the intratracheal method is nonphysiologic and it is a relatively invasive method [99].

The nucleic acid siRNA can be subjected to chemical modifications of the sugar, backbone, or the bases of ribonucleotides in order to avoid nucleases [100]. Instead of nucleic acid modification, naked siRNA should be delivered with a nanocarrier. Viral carriers are very efficient in delivering DNA but on the contrary due to their side effects and uncontrolled replication, they are not appropriate for clinical applications [101].

Nonviral carriers used are lipids and liposomes, such as Lipofectamine 2000™, which are popular in vitro transfection reagent. Invivofectamine™ is a lipid nanoparticle designed for a high-efficiency in vivo delivering of siRNA or miRNA, polymers such as atelocollagen, chitosan, a natural polysaccharide accepted as biocompatible and biodegradable, and peptides.

Since the first applications of antisense oligonucleotides in respiratory diseases in 2003, numerous studies have been reported and several patents have been developed based on RNAi strategies in asthma and allergy [102]. Recently several studies have been reported of local siRNA delivery in experimental model of asthma. Asai-Tajiri Y et al. studied the role of CD86 on dendritic cells in the reactivation of allergen-specific Th2. They performed an intratracheal siRNA against CD86 of bone marrow-derived dendritic cells (BMDCs) in vitro and on a murine experimental model of asthma. The results show that local administration of CD86 siRNA suppresses airway inflammation and hyperresponsiveness in a model of asthma [89].

In 2014, Wu W et al. investigated whether silencing of c-kit with intranasal siRNA could reduce inflammation in allergic asthma. A mouse model of experimental asthma was treated with intranasal administration of anti-c-kit siRNA nanoparticles to inhibit the c-kit expression. Local administration of siRNA effectively inhibited c-kit expression and decreased the airway mucus secretion and infiltration of eosinophils in broncho alveolar lavage fluid (BALF). Moreover IL-4 and IL-5 were reduced. Authors conclude that intranasally delivered siRNA nanoparticles, targeting c-kit, can decrease the inflammatory response in experimental allergic asthma [103]. The *CKIT* gene was previously silenced in a murine model of experimental asthma by systemic administration observing the same effects over the gene as by the local administration [104].

Chen et al. developed a siRNA against NGF to attenuate airway hyperresponsiveness and further elucidate the underlying mechanism. In a murine allergic asthma model, the siNGF into the lentiviral expression system was intratracheally delivered in ovalbumin-sensitized mice. As a positive control, an inhibitor targeting NFG receptor was used. The knockdown NGF, in pulmonary epithelium, significantly reduced airway resistance in vivo. Levels of NGF, proinflammatory cytokines, and infiltrated eosinophils in airway were decreased in the siRNA group. The findings suggested that NGF is a target for a potential therapy of antigen-induced airway hyperresponsiveness via attenuation of the airway innervations and inflammation in asthma [105].

The strategies directed to the manipulation of control gene expression for the treatment or prevention of diseases are likely to play a crucial role in the future of clinical therapies. The academia and the clinic should be closer than before. To have a real translation from bench to bead, academic, clinical researchers, and pharmaceutical industry must have the same focus.

Often academic research is more concerned about publishable results that are not always clinically relevant. Many studies focus on the design of siRNA molecules instead of on the delivery perspective. In a siRNA therapy approach for treating human lung diseases as asthma and allergy, the mechanism of delivery should be via inhalation of nonviral nanoparticles. Well-designed studies are necessary to have a deep knowledge of RNAi and to get this technology to expand.

Acknowledgments

This work was supported by grants of the Junta de Castilla y León ref. GRS1047/A/14, GRS1189/A/15, and BIO/SA73/15; and by the project "Efecto del Ácido Retinóico en la enfermedad alérgica. Estudio transcripcional y su traslación a la clínica", PI13/00564, integrated into the "Plan Estatal de I+D+I 2013–2016" and cofunded by the "ISCIII-Subdirección General de Evaluación y Fomento de la investigación" and the European Regional Development Fund (FEDER).

References

1. Lee TI, Young RA (2000) Transcription of eukaryotic protein-coding genes. Annu Rev Genet 34:77–137
2. Lemon B, Tjian R (2000) Orchestrated response: a symphony of transcription factors for gene control. Genes Dev 14:2551–2569
3. Karin M (1999) Too many transcription factors: positive and negative interactions. New Biol 2:126–131
4. Latchman DS (1997) Transcription factors: an overview. Int J Biochem Cell Biol 29: 1305–1312
5. Nikolov DB, Burley SK (1997) RNA polymerase II transcription initiation: a structural view. Proc Natl Acad Sci U S A 94:15–22
6. Roeder RG (1996) The role of general initiation factors in transcription by RNA polymerase II. Trends Biochem Sci 21:327–335
7. Mitchell PJ, Tjian R (1989) Transcriptional regulation in mammalian cells by sequence-specific DNA binding proteins. Science 245:371–378
8. Ptashne M, Gann A (1997) Transcriptional activation by recruitment. Nature 386: 569–577
9. Kouzarides T (2007) Chromatin modifications and their function. Cell 128:693–705
10. Bell JT, Pai AA, Pickrell JK et al (2011) DNA methylation patterns associate with genetic and gene expression variation in HapMap cell lines. Genome Biol 12:R10
11. Bird A (2007) Perceptions of epigenetics. Nature 447:396–398
12. Goldberg AD, Allis CD, Bernstein E (2007) Epigenetics: a landscape takes shape. Cell 128:635–638

13. Robertson KD (2005) DNA methylation and human disease. Nat Rev Genet 6:597–610
14. Strahl BD, Allis CD (2000) The language of covalent histone modifications. Nature 403(6765):41–45
15. Black DL (2003) Mechanisms of alternative pre-messenger RNA splicing. Annu Rev Biochem 72:291–336
16. Voet D, Voet JG, Pratt CW (2006) Fundamentals of biochemistry: life at the molecular level. Wiley, Hoboken NJ
17. Khoury GA, Baliban RC, Floudas CA (2011) Proteome-wide post-translational modification statistics: frequency analysis and curation of the swiss-prot database. Sci Rep 1:90
18. Towbin H, Staehelin T, Gordon J (1979) Electrophoretic transfer of proteins from polyacrylamide gels to nitrocellulose sheets: procedure and some applications. Proc Natl Acad Sci U S A 76:4350–4354
19. Burnette WN (1981) "Western blotting": electrophoretic transfer of proteins from sodium dodecyl sulfate--polyacrylamide gels to unmodified nitrocellulose and radiographic detection with antibody and radioiodinated protein A. Anal Biochem 112:195–203
20. Southern EM (1975) Detection of specific sequences among DNA fragments separated by gel electrophoresis. J Mol Biol 98:503–517
21. Alwine JC, Kemp DJ, Stark GR (1977) Method for detection of specific RNAs in agarose gels by transfer to diazobenzyloxymethyl-paper and hybridization with DNA probes. Proc Natl Acad Sci U S A 74:5350–5354
22. Gershoni JM (1988) Protein blotting: a manual. Methods Biochem Anal 33:1–58
23. Kurien BT, Scofield RH (2006) Western blotting. Methods 38:283–293
24. Legocki RP, Verma DP (1981) Multiple immunoreplica technique: screening for specific proteins with a series of different antibodies using one polyacrylamide gel. Anal Biochem 111:385–392
25. Otter T, King SM, Witman GB (1987) A two-step procedure for efficient electrotransfer of both high-molecular-weight (greater than 400,000) and low-molecular-weight (less than 20,000) proteins. Anal Biochem 162:370–377
26. Leong MM, Milstein C, Pannell R (1986) Luminescent detection method for immunodot, Western, and Southern blots. J Histochem Cytochem 34:1645–1650
27. Vachereau A (1989) Luminescent immunodetection of western-blotted proteins from coomassie-stained polyacrylamide gel. Anal Biochem 179:206–208
28. Sutton R, Skerritt JH, Baldo BA (1984) The diversity of allergens involved in bakers' asthma. Clin Allergy 14:93–107
29. Weiss W, Vogelmeier C, Gorg A (1993) Electrophoretic characterization of wheat grain allergens from different cultivars involved in bakers' asthma. Electrophoresis 14:805–816
30. Suphioglu C, Singh MB, Taylor P et al (1992) Mechanism of grass-pollen-induced asthma. Lancet 339:569–572
31. Haitchi HM, Powell RM, Shaw TJ et al (2005) ADAM33 expression in asthmatic airways and human embryonic lungs. Am J Respir Crit Care Med 171:958–965
32. Sadowski HB, Shuai K, Darnell JE Jr et al (1993) A common nuclear signal transduction pathway activated by growth factor and cytokine receptors. Science 261:1739–1744
33. Zhong Z, Wen Z, Darnell JE Jr (1994) Stat3: a STAT family member activated by tyrosine phosphorylation in response to epidermal growth factor and interleukin-6. Science 264:95–98
34. Chen Z, Wang S, Erekosima N et al (2013) IL-4 confers resistance to IL-27-mediated suppression on CD4+ T cells by impairing signal transducer and activator of transcription 1 signaling. J Allergy Clin Immunol 132:912–921
35. Xie M, Mustovich AT, Jiang Y et al (2015) IL-27 and type 2 immunity in asthmatic patients: association with severity, CXCL9, and signal transducer and activator of transcription signaling. J Allergy Clin Immunol 135:386–394
36. Mazzeo C, Canas JA, Zafra MP et al (2015) Exosome secretion by eosinophils: a possible role in asthma pathogenesis. J Allergy Clin Immunol 135:1603–1613
37. Garner MM, Revzin A (1981) A gel electrophoresis method for quantifying the binding of proteins to specific DNA regions: application to components of the Escherichia coli lactose operon regulatory system. Nucleic Acids Res 9:3047–3060
38. Fried M, Crothers DM (1981) Equilibria and kinetics of lac repressor-operator interactions by polyacrylamide gel electrophoresis. Nucleic Acids Res 9:6505–6525
39. Chelm BK, Geiduschek EP (1979) Gel electrophoretic separation of transcription complexes: an assay for RNA polymerase selectivity and a method for promoter mapping. Nucleic Acids Res 7:1851–1867
40. Eisinger J (1971) Visible gel electrophoresis and the determination of association con-

stants. Biochem Biophys Res Commun 44: 1135–1142

41. Hellman LM, Fried MG (2007) Electrophoretic mobility shift assay (EMSA) for detecting protein-nucleic acid interactions. Nat Protoc 2:1849–1861
42. Varshavsky AJ, Bakayev VV, Georgiev GP (1976) Heterogeneity of chromatin subunits *in vitro* and location of histone H1. Nucleic Acids Res 3:477–492
43. Fried MG (1989) Measurement of protein-DNA interaction parameters by electrophoresis mobility shift assay. Electrophoresis 10: 366–376
44. Fried MG, Crothers DM (1984) Kinetics and mechanism in the reaction of gene regulatory proteins with DNA. J Mol Biol 172: 263–282
45. Mori A, Kaminuma O, Suko M et al (1997) Two distinct pathways of interleukin-5 synthesis in allergen-specific human T-cell clones are suppressed by glucocorticoids. Blood 89:2891–2900
46. In KH, Silverman ES, Asano K et al (1999) Mutations in the human 5-lipoxygenase gene. Clin Rev Allergy Immunol 17:59–69
47. Papi A, Papadopoulos NG, Stanciu LA et al (2001) Effect of desloratadine and loratadine on rhinovirus-induced intercellular adhesion molecule 1 upregulation and promoter activation in respiratory epithelial cells. J Allergy Clin Immunol 108:221–228
48. Oguma T, Palmer LJ, Birben E et al (2004) Role of prostanoid DP receptor variants in susceptibility to asthma. N Engl J Med 351:1752–1763
49. Isidoro-Garcia M, Sanz C, Garcia-Solaesa V et al (2011) PTGDR gene in asthma: a functional, genetic, and epigenetic study. Allergy 66:1553–1562
50. Qiu R, Yang Y, Zhao H et al (2013) Signal transducer and activator of transcription 6 directly regulates human ORMDL3 expression. FEBS J 280:2014–2026
51. Zhuang LL, Jin R, Zhu LH et al (2013) Promoter characterization and role of cAMP/PKA/CREB in the basal transcription of the mouse ORMDL3 gene. PLoS One 8:e60630
52. Hwang SS, Kim YU, Lee S et al (2013) Transcription factor YY1 is essential for regulation of the Th2 cytokine locus and for Th2 cell differentiation. Proc Natl Acad Sci U S A 110:276–281
53. Ren B, Robert F, Wyrick JJ et al (2000) Genome-wide location and function of DNA binding proteins. Science 290: 2306–2309
54. Spencer VA, Sun JM, Li L, Davie JR (2003) Chromatin immunoprecipitation: a tool for studying histone acetylation and transcription factor binding. Methods 31:67–75
55. Weinmann AS, Bartley SM, Zhang T et al (2001) Use of chromatin immunoprecipitation to clone novel E2F target promoters. Mol Cell Biol 21:6820–6832
56. Weinmann AS, Farnham PJ (2002) Identification of unknown target genes of human transcription factors using chromatin immunoprecipitation. Methods 26:37–47
57. Weinmann AS, Yan PS, Oberley MJ et al (2002) Isolating human transcription factor targets by coupling chromatin immunoprecipitation and CpG island microarray analysis. Genes Dev 16:235–244
58. Yan Y, Kluz T, Zhang P et al (2003) Analysis of specific lysine histone H3 and H4 acetylation and methylation status in clones of cells with a gene silenced by nickel exposure. Toxicol Appl Pharmacol 190:272–277
59. Gilmour DS, Lis JT (1984) Detecting protein-DNA interactions in vivo: distribution of RNA polymerase on specific bacterial genes. Proc Natl Acad Sci U S A 81: 4275–4279
60. Gilmour DS, Lis JT (1985) In vivo interactions of RNA polymerase II with genes of Drosophila melanogaster. Mol Cell Biol 5:2009–2018
61. Gilmour DS, Lis JT (1986) RNA polymerase II interacts with the promoter region of the noninduced hsp70 gene in Drosophila melanogaster cells. Mol Cell Biol 6:3984–3989
62. Solomon MJ, Varshavsky A (1985) Formaldehyde-mediated DNA-protein crosslinking: a probe for in vivo chromatin structures. Proc Natl Acad Sci U S A 82: 6470–6474
63. Wells J, Farnham PJ (2002) Characterizing transcription factor binding sites using formaldehyde crosslinking and immunoprecipitation. Methods 26:48–56
64. Mardis ER (2007) ChIP-seq: welcome to the new frontier. Nat Methods 4:613–614
65. Mardis ER (2008) Next-generation DNA sequencing methods. Annu Rev Genomics Hum Genet 9:387–402
66. Robertson G, Hirst M, Bainbridge M et al (2007) Genome-wide profiles of STAT1 DNA association using chromatin immunoprecipitation and massively parallel sequencing. Nat Methods 4:651–657
67. Knight JC, Keating BJ, Kwiatkowski DP (2004) Allele-specific repression of

lymphotoxin-alpha by activated B cell factor-1. Nat Genet 36:394–399
68. Knight JC, Keating BJ, Rockett KA et al (2003) In vivo characterization of regulatory polymorphisms by allele-specific quantification of RNA polymerase loading. Nat Genet 33:469–475
69. Xiang Y, Qin XQ, Liu HJ et al (2012) Identification of transcription factors regulating CTNNAL1 expression in human bronchial epithelial cells. PLoS One 7:e31158
70. Malhotra D, Portales-Casamar E, Singh A et al (2010) Global mapping of binding sites for Nrf2 identifies novel targets in cell survival response through ChIP-Seq profiling and network analysis. Nucleic Acids Res 38: 5718–5734
71. Sharma S, Kho AT, Chhabra D et al (2014) Glucocorticoid genes and the developmental origins of asthma susceptibility and treatment response. Am J Respir Cell Mol Biol 52: 543–553
72. Luca F, Maranville JC, Richards AL et al (2013) Genetic, functional and molecular features of glucocorticoid receptor binding. PLoS One 8:e61654
73. Valapour M, Guo J, Schroeder JT et al (2002) Histone deacetylation inhibits IL4 gene expression in T cells. J Allergy Clin Immunol 109:238–245
74. Cui ZL, Gu W, Ding T, Peng XH et al (2013) Histone modifications of Notch1 promoter affect lung CD4+ T cell differentiation in asthmatic rats. Int J Immunopathol Pharmacol 26:371–381
75. Bernstein E, Caudy AA, Hammond SM et al (2001) Role for a bidentate ribonuclease in the initiation step of RNA interference. Nature 409:363–366
76. Fire A, Xu S, Montgomery MK, Kostas SA et al (1998) Potent and specific genetic interference by double-stranded RNA in Caenorhabditis elegans. Nature 391: 806–811
77. Elbashir SM, Lendeckel W, Tuschl T (2001) RNA interference is mediated by 21- and 22-nucleotide RNAs. Genes Dev 15:188–200
78. Hammond SM, Bernstein E, Beach D et al (2000) An RNA-directed nuclease mediates post-transcriptional gene silencing in Drosophila cells. Nature 404:293–296
79. Martinez J, Tuschl T (2004) RISC is a 5′ phosphomonoester-producing RNA endonuclease. Genes Dev 18:975–980
80. Lecellier CH, Dunoyer P, Arar K et al (2005) A cellular microRNA mediates antiviral defense in human cells. Science 308: 557–560
81. Gitlin L, Karelsky S, Andino R (2002) Short interfering RNA confers intracellular antiviral immunity in human cells. Nature 418: 430–434
82. Gitlin L, Andino R (2003) Nucleic acid-based immune system: the antiviral potential of mammalian RNA silencing. J Virol 77: 7159–7165
83. Sijen T, Plasterk RH (2003) Transposon silencing in the Caenorhabditis elegans germ line by natural RNAi. Nature 426:310–314
84. Aigner A (2006) Gene silencing through RNA interference (RNAi) in vivo: strategies based on the direct application of siRNAs. J Biotechnol 124:12–25
85. Aigner A (2007) Applications of RNA interference: current state and prospects for siRNA-based strategies in vivo. Appl Microbiol Biotechnol 76:9–21
86. Rao DD, Senzer N, Cleary MA et al (2009) Comparative assessment of siRNA and shRNA off target effects: what is slowing clinical development. Cancer Gene Ther 16: 807–809
87. Li L, Yang Y, Stevens RL (2003) RasGRP4 regulates the expression of prostaglandin D2 in human and rat mast cell lines. J Biol Chem 278:4725–4729
88. Al-Alwan LA, Chang Y, Mogas A et al (2013) Differential roles of CXCL2 and CXCL3 and their receptors in regulating normal and asthmatic airway smooth muscle cell migration. J Immunol 191:2731–2741
89. Asai-Tajiri Y, Matsumoto K, Fukuyama S et al (2014) Small interfering RNA against CD86 during allergen challenge blocks experimental allergic asthma. Respir Res 15:132
90. Hong GU, Park BS, Park JW et al (2013) IgE production in CD40/CD40L cross-talk of B and mast cells and mediator release via TGase 2 in mouse allergic asthma. Cell Signal 25:1514–1525
91. Papi A, Contoli M, Adcock IM et al (2013) Rhinovirus infection causes steroid resistance in airway epithelium through nuclear factor kappaB and c-Jun N-terminal kinase activation. J Allergy Clin Immunol 132:1075–1085.e6
92. Sundaram K, Mitra S, Gavrilin MA et al (2015) House dust mite allergens induce monocyte IL-1beta production triggering an IkappaBzeta dependent GMCSF release from human lung epithelial cells. Am J Respir Cell Mol Biol 53:400–411

93. Trian T, Girodet PO, Ousova O et al (2006) RNA interference decreases PAR-2 expression and function in human airway smooth muscle cells. Am J Respir Cell Mol Biol 34:49–55
94. Zhou J, Herring BP (2005) Mechanisms responsible for the promoter-specific effects of myocardin. J Biol Chem 280:10861–10869
95. Haj-Salem I, Fakhfakh R, Berube JC et al (2015) MicroRNA-19a enhances proliferation of bronchial epithelial cells by targeting TGFbetaR2 gene in severe asthma. Allergy 70:212–219
96. Garbacki N, Di Valentin E, Huynh-Thu VA et al (2011) MicroRNAs profiling in murine models of acute and chronic asthma: a relationship with mRNAs targets. PLoS One 6:e16509
97. Williams AE, Larner-Svensson H et al (2009) MicroRNA expression profiling in mild asthmatic human airways and effect of corticosteroid therapy. PLoS One 4:e5889
98. Jardim MJ, Dailey L, Silbajoris R et al (2012) Distinct microRNA expression in human airway cells of asthmatic donors identifies a novel asthma-associated gene. Am J Respir Cell Mol Biol 47:536–542
99. Lam JK, Liang W, Chan HK (2012) Pulmonary delivery of therapeutic siRNA. Adv Drug Deliv Rev 64:1–15
100. Merkel OM, Rubinstein I, Kissel T (2014) siRNA delivery to the lung: what's new? Adv Drug Deliv Rev 75:112–128
101. Thomas CE, Ehrhardt A, Kay MA (2003) Progress and problems with the use of viral vectors for gene therapy. Nat Rev Genet 4:346–358
102. Isidoro-Garcia M, Davila I (2007) Therapeutic applications of antisense oligonucleotides in asthma and allergy. Recent Pat Inflamm Allergy Drug Discov 1:171–175
103. Wu W, Chen H, Li YM et al (2014) Intranasal sirna targeting c-kit reduces airway inflammation in experimental allergic asthma. Int J Clin Exp Pathol 7:5505–5514
104. Wu W, Wang T, Dong JJ et al (2012) Silencing of c-kit with small interference RNA attenuates inflammation in a murine model of allergic asthma. Int J Mol Med 30:63–68
105. Chen YL, Huang HY, Lee CC et al (2014) Small interfering RNA targeting nerve growth factor alleviates allergic airway hyperresponsiveness. Mol Ther Nucleic Acids 3:e158

Chapter 7

Interactions of DNA and Proteins: Electrophoretic Mobility Shift Assay in Asthma

Virginia García-Solaesa and Catalina S. Sanz-Lozano

Abstract

Electrophoretic mobility shift assays (EMSA) are used to characterize interactions between nucleic acids and proteins in native conditions. This is based on the fact that the electrophoretic mobility of a nucleic acid becomes slower when it forms complexes with proteins. There are many different variants and applications of this methodology. In this chapter we describe a detailed EMSA protocol applied to the study of asthma.

Key words Binding, Electrophoretic mobility shift assay, EMSA, Gel retardation, Protein–DNA complex

1 Introduction

Many genes have been associated with the pathogenesis of asthma. Different strategies such as linkage analysis and association studies with candidate genes have been used to analyze genetic factors of asthma. Methodological differences, in terms of levels of significance, statistical power, definition of phenotypes, criteria of selection, or sample size, limit the replication of data. In addition, the confirmation of the functional effects of the candidate genes associated with asthma is essential.

Electrophoretic mobility shift assays (EMSA) are carried out to study interactions between nucleic acids and proteins, to identify sequence-specific DNA-binding proteins (such as transcription factors). This allows analyzing the role played by promoter mutations as it has been described for the promoter region of the *PTGDR* gene. Some variants in the *PTGDR* promoter modify the union affinity of certain transcriptional factors, which leads to changes in the gene expression patterns with possible consequences on the development of asthma [1, 2].

This methodology is based on the different migration of DNA: protein complexes in native (non-denaturing) electrophoresis compared to the free DNA fragments [3] (Fig. 1). The DNA

María Isidoro-García (ed.), *Molecular Genetics of Asthma*, Methods in Molecular Biology, vol. 1434,
DOI 10.1007/978-1-4939-3652-6_7, © Springer Science+Business Media New York 2016

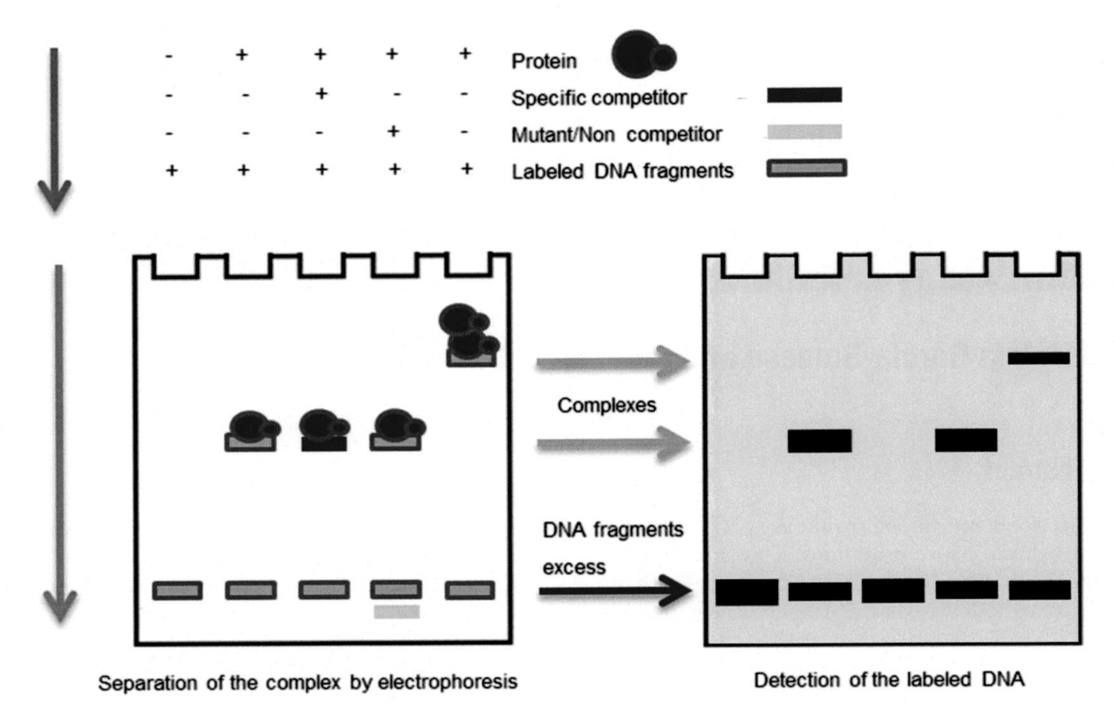

Fig. 1 Electrophoretic mobility shift assay basis [4]

migration is shifted or retarded when bound to protein, the so-called gel shift or gel retardation [4]. The ability to detect DNA:protein complexes lies in their stability in each step of the assay. In electrophoresis, the gel matrix promotes the stabilization of the DNA:protein interaction [5]. In addition, due to the low ionic strength of the electrophoresis buffer, the transient or very labile interactions are kept and can be detected.

However, the technique may require competitors to establish the specificity of the interactions detected. When purified proteins are used it is not necessary to add a nonspecific DNA competitor, while with total protein extracts it is absolutely essential adding a nonspecific DNA competitor to the union reaction. It has to be taken into account that some reactions may require different reaction temperatures.

2 Materials

2.1 DNA Fragments and Protein Extract

1. TEN buffer: 10 mM Tris, 1 mM EDTA, 0.1 M NaCl, pH 8.0.
2. Distilled water free of nucleases.
3. Thermocycler.
4. Double-stranded DNA fragments (ds-DNA) obtained through the binding of their respective simple chain oligonucleotides (ss-oligonucleotides) (*see* **Note 1**).

5. 0.1 μg/μl Control ds-oligonucleotide 39mer unlabeled.
6. Commercial extract of nuclear proteins derived from leukocytes of a healthy-individual peripheral blood (AMS Biotechnology). The concentration is 1 mg/ml and the buffer contains HEPES, pH 7.9, glycerol, $MgCl_2$, NaCl, EDTA, and a set of protease inhibitors (*see* **Note 2**).
7. Positive control factor Oct2A: 25–75 ng/μl in 30 mM Tris–HCl, pH 8.0, 0.2 mM EDTA, 2 mM DTT, 0.2 M NaCl (DIG Gel Shift Kit 2nd Generation, Roche Applied Science).

2.2 Labeling of DNA Fragments

1. DIG Gel Shift Kit 2nd Generation (Roche Applied Science) containing:
   - 5× Labeling buffer: 1 M Potassium cacodylate, 0.125 M Tris–HCl, 1.25 mg/ml bovine serum albumin, pH 6.6.
   - 25 mM $CoCl_2$ solution.
   - DIG-ddUTP solution: 1 mM Digoxigenin-11-ddUTP in double-distilled water.
   - Terminal transferase recombinant: 400 U/μl Terminal transferase recombinant, 60 mM K-phosphate, pH 7.2 at 4 °C, 150 mM KCl, 1 mM 2-mercaptoethanol, 0.5 % Triton X-100, 50 % glycerol.
2. TEN buffer: 10 mM Tris, 1 mM EDTA, 0.1 M NaCl, pH 8.0.
3. 0.2 M EDTA: 0.2 M Ethylenediaminotetraacetic acid, pH 8.0.
4. Distilled water free of nucleases.

2.3 Purification of Labeled Oligonucleotides

1. Illustra Probe Quant G-50 Micro columns (GE Healthcare).
   - Probe buffer: 10 mM Tris–HCl, pH 8, 0 1 mM EDTA, and 150 mM NaCl.
   - Sepharose micro columns: *Prepacked Sephadex*™ *G-50 DNA Grade* in column storage buffer: 150 mM STE, 0.15 % *Kathon CG/ICP Biocide* as preservative.

2.4 Binding Reaction

1. DIG Gel Shift Kit 2nd Generation (Roche Applied Science):
   - 5× Binding buffer: 100 mM Hepes, pH 7.6, 5 mM EDTA, 50 mM $(NH_4)_2$ SO_4, 5 mM DTT, Tween 20, 1 %(w/v), 150 mM KCl.
   - 1 μg/μl Poly [d(I-C)].
   - 1 μg/μl Poly [d(A-T)].
   - 0.1 μg/μl Poly L-lysine.

2.5 Non-denaturing Polyacrylamide Gel Electrophoresis

1. *N,N,N′,N′*-tetramethylethylenediamine (TEMED).
2. 10 % (w/v) Ammonium persulfate (APS).
3. 40 % (w/v) Acrylamide/Bisacrylamide 29/1.

4. 10× TBE buffer: 890 mM Tris-borate, 890 mM boric acid, 20 mM EDTA, pH 8.2–8.4.
5. Loading buffer with bromophenol blue: 0.25× TBE buffer, 60 %; glycerol, 40 %; bromophenol blue, 0.2 % (w/v).

2.6 Blotting and Cross-Linking

1. Nylon membranes, positively charged 0.45 μm (Roche Applied Science) (*see* **Note 3**).
2. Whatman 3 MM paper.
3. Electro-blotting device (PowerPac Basic Bio-Rad).
4. 10× TBE buffer: 890 mM Tris-borate, 890 mM boric acid, 20 mM EDTA, pH 8.2–8.4.
5. UV-cross linker (BIO-LINK® BLX-E254, Vilber Lourmat).
6. 10× SSC blotting buffer.

2.7 Chemi-luminescent Detection

1. DIG Gel Shift Kit 2nd Generation (Roche Applied Science) containing:
   - Anti-digoxigenin-AP: 750 U/ml Polyclonal sheep, anti-digoxigenin, Fab fragments conjugated with alkaline phosphatase.
   - 10 mg/ml of CSPD(3-(4-methoxyspiro[8) decan]-4-yl) phenyl phosphate).
2. DIG Wash and Block buffer set (Roche Applied Science) containing:
   - 10× Washing buffer: 10× Maleic acid buffer with 3–5 % Tween 20 (v/v).
   - 10× Maleic acid buffer.
   - Blocking solution.
   - 10× Detection buffer: 1 M Tris–HCl, pH 9.5, 1 M NaCl.
   - High-sensitive autoradiographic film (Kodak).
   - "Kodak *processing chemicals for autoradiography films*" (Sigma-Aldrich Co. LLC, USA).

3 Methods

3.1 DNA Fragments

The study of the genetic variant effect over the transcription factor binding requires the synthesis of two pairs of ss-oligonucleotides, one pair containing the sequence with the conserved position and the other with the mutant base. These are designed from the reference sequence deposited in GenBank database (http://www2.ncbi.nlm.nih.gov/cgi-bin/genbank), and commercially acquired. Each pair consisted of a complementary forward and reverse primer. The size of the resulting ds-DNA fragments should be between 30 and 40 bp (*see* **Note 1**). In addition, these ds-DNA

Fig. 2 Double-stranded oligonucleotide (ds-oligonucleotide). The union is outlined by complementarity of the ddUTP (joined the digoxigenin) to the free adenines at 3′ ends through the action of terminal transferase

```
5´-GTACGGAGTATCCAGCTCCGTAGCATGCAAATCCTCTGG      - 3´
3´-      CCTCATAGGTCGAGGCATCGTACGTTTAGGAGACCAGCT- 5´
```

Fig. 3 Sequence of control ds-oligonucleotide 39mer. The binding site of the protein Oct2A is shown in *pink*

fragments should have 3′ ss ends with some free adenines (A) in their sequence, allowing the complementary chain to be filled and labeled with DIG-11-ddUTP when the terminal transferase is added (Fig. 2) [6]. The ss-oligonucleotides are supplied as a lyophilized and need to be reconstituted to 100 μM concentration with distilled water free of nucleases.

The synthesis of ds-DNA fragments is carried out by the annealing of each ss-oligonucleotide with its complementary. Two reactions are made with each partner at the concentrations of 4 and 100 μM (dilution in TEN buffer).

1. Prepare equimolar mixtures of each ss-oligonucleotides pair, incubate in a thermocycler for 10 min at 95 °C, and then apply a temperature ramp to get a slow cooling to 20 °C.
2. Once annealed, double-stranded oligonucleotides (ds-oligonucleotides) have to be stored at −20 °C. It is possible to continue with the labeling process at 4 μM concentration.

3.2 Labeling of Oligonucleotides

The labeling is carried out with DIG-11-ddUTP, a digoxigenin (DIG) molecule attached at the C5 position of uridine to dideoxy-uracil-triphosphate (ddUTP) using a spacer arm of 11 carbon atoms. The digoxigenin is a steroid that comes from the leaves and flowers of *Digitalis purpurea* and *Digitalis lanata*, the natural source of this compound. Anti-digoxigenin antibodies do not recognize nor bind to any other biological material [6]. The labeling of the oligonucleotides occurs through the action of a terminal transferase recombinant that catalyzes the union of the 11-DIG-ddUTP in the 3′ ends of the DNA fragments. The labeling of the ds-oligonucleotides is held in a final volume of 20 μl with the reagents contained in the DIG Gel Shift Kit 2nd Generation [6]. An oligonucleotide containing the Oct2A protein-binding site in its sequence (Fig. 3) is used as positive control. DIG Gel Shift Kit 2nd Generation provides the DNA fragment labeled with DIG-ddUTP and unlabeled. The unlabeled control oligonucleotide is used to assess the gel shift assay at each step.

1. For each reaction, add 4 μl of Labeling buffer, 4 μl of $CoCl_2$ solution, 1 μl of DIG-ddUTP solution, 1 μl of terminal transferase recombinant, and 40 pmol of ds-oligonucleotide or 4 pmol for control oligonucleotide (*see* **Note 4**).
2. Incubate the mixture at 37 °C for 15 min within the thermocycler.
3. After this time stop the reaction by adding 2 μl of 0.2 M EDTA.
4. Finally, add 3 μl of free of nuclease distilled water to achieve a final concentration of labeled ds-oligonucleotide of 1.6 pmol/μl and 0.16 pmol/μl for the control oligonucleotide.

3.3 Purification of Labeled Oligonucleotides

Filtered on a column of sepharose, the labeled DNA fragments larger than 20 pb are broken away from the excess of DIG-ddUTP. Molecules larger than the pores of this resin are excluded from the gel matrix as it is the case of the labeled ds-oligonucleotides, while those of smaller size, the leftover labeled nucleotides, are retained in the column and therefore do not appear in the elute. The purification of labeled ds-oligonucleotides is held in a final volume of 50 μl with Illustra Probe Quant G-50 Micro columns.

1. Dilute labeled ds-oligonucleotides samples (25 μl) to a volume of 50 μl with probe buffer type 1.
2. Resuspend the sepharose resin contained in the micro column by vortex.
3. Break the terminal end of the column and remove the buffer centrifuging at 735 × *g* for 1 min (*see* **Note 5**).
4. Place the column in a fresh DNase-free 1.5 ml microcentrifuge tube and apply the sample on the resin.
5. Spin the columns for 2 min at 735 × *g* and collect the eluents with the labeled and purified ds-oligonucleotides.
6. Identify the purified samples with the date of labeling and store at −20 °C. The stability of the labeling at these conditions is 7–9 days.

3.4 Binding Reaction

The binding conditions, such as salt concentration and pH, can determine the interactions between DNA and proteins, so it may be necessary to add Mg2+, Zn2+, Ca2+, detergent, or spermidine depending on the samples. In the gel shift assays, the use of poly L-lysine allows the optimization of specific DNA:protein complex formation, because simple peptides can increase the apparent affinity of the DNA binding.

Poly [d(I-C)] is used as nonspecific competitor if the fragment were rich in guanine (G) and cytosine (C) and poly [d(A-T)] when the fragment were rich in adenine (A) and thymine (T).

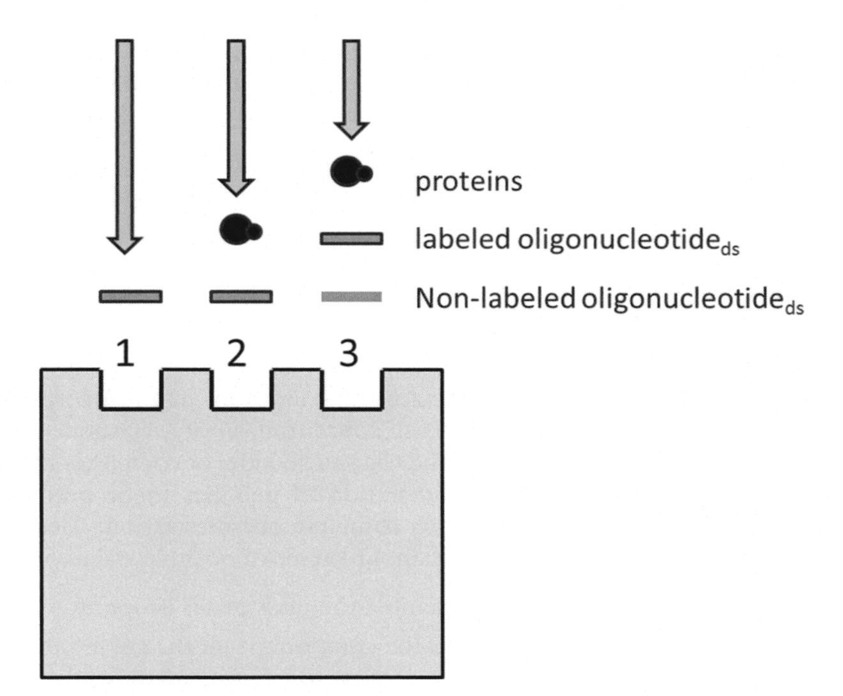

Fig. 4 Gel shift assay scheme. Representation of three binding reaction components that will be analyzed by electrophoresis

The specific competitor used in each experiment is the same ds-oligonucleotide unlabeled and at a 100 μM concentration. It is possible to improve the formation of specific DNA:protein complexes by using albumin. Similarly, the order of the addition of labeled oligonucleotide and nuclear extract can substantially influence the specificity of DNA-protein interactions [6].

Each gel shift assay implies three reactions: the first one with the labeled ds-oligonucleotide, the second one with the labeled ds-oligonucleotide and protein extract, and the third with the labeled ds-oligonucleotide, the protein extract, and the competitor ds-oligonucleotide (Fig. 4). By comparing the results of these three reactions it could be interpreted whether such DNA-protein/s complexes are formed and whether the bindings are specific.

1. Prepare a properly identified Eppendorf tube for each reaction, in a final volume of 20 μl (adjust with distilled water).
   - 4 μl of binding buffer; final concentration: 20 mM HEPES, pH 7.6, 1 mM EDTA, 10 mM $(NH_4)_2SO_4$ 1 mM DTT, 0.2% Tween 20 (w/v), 30 mM KCl.
   - 1 μl Poly [d(I-C)] (final concentration: 0.05 μg/μl).
   - 1 μl Poly-L-lysine (final concentration: 0.005 μg/μl).
   - 2 μl of ds-oligonucleotide (labeled and purified).

- 2 μl of protein extract or 1 μl of Oct2A factor in the control reaction (only in reactions 2 and 3).
- 1 μl of competitor ds-oligonucleotide (only in reaction 3) (*see* **Note 6**).

2. Mix the reaction by vortex and spin.
3. Incubate for 30 min at room temperature (25 °C) and subsequently put in ice before being applied in the electrophoresis gel (*see* **Note 7**).

3.5 Non-denaturing Gel Electrophoresis

3.5.1 Polyacrylamide Gel

The DNA:protein complexes are separated by electrophoresis on non-denaturing conditions in a polyacrylamide gel and 0.5× TBE buffer. The concentration of acrylamide in the gel depends on the size of the oligonucleotides or complexes with protein/s (*see* **Note 8**). The preparation of gels has to be performed the day before to ensure its complete polymerization. Clean the crystals with 100% pure ethanol to remove possible traces of polyacrylamide.

1. Assemble the glass plates properly.
2. Mix the components of the gel in the order showed in Table 1. The total volume needed depends on the gel mold and the manufacturer instructions. Mixtures are prepared under shaking and cold conditions (4 °C) in order to avoid early polymerization of polyacrylamide, and add the APS and TEMED at the end. The APS is used as a catalyst for the co-polymerization of bisacrylamide and acrylamide gels. The polymerization reaction is caused by the free radicals generated in oxidation-reduction reactions in which TEMED participates as a co-catalyst of APS (Fig. 5) [7]. The polymerization begins as soon as the TEMED is added to the mixture. The lowest possible concentration of catalysts must be used that allows polymerization in optimal time.

Table 1
Volume of components required for preparing different concentrations of polyacrylamide gels

	4.1% (200 bp)	8%	10%	12%
Distilled water	16.2 ml	13.6 ml	12.25 ml	10.9 ml
29/1 Acrylamide/bisacrylamide	2.72 ml	5.32 ml	6.65 ml	8 ml
10× TBE buffer	1 ml	1 ml	1 ml	1 ml
10% Ammonium persulfate (APS)	20 μl	20 μl	20 μl	20 μl
TEMED	80 μl	80 μl	80 μl	80 μl
Total volume (ml)	20	20	20	20

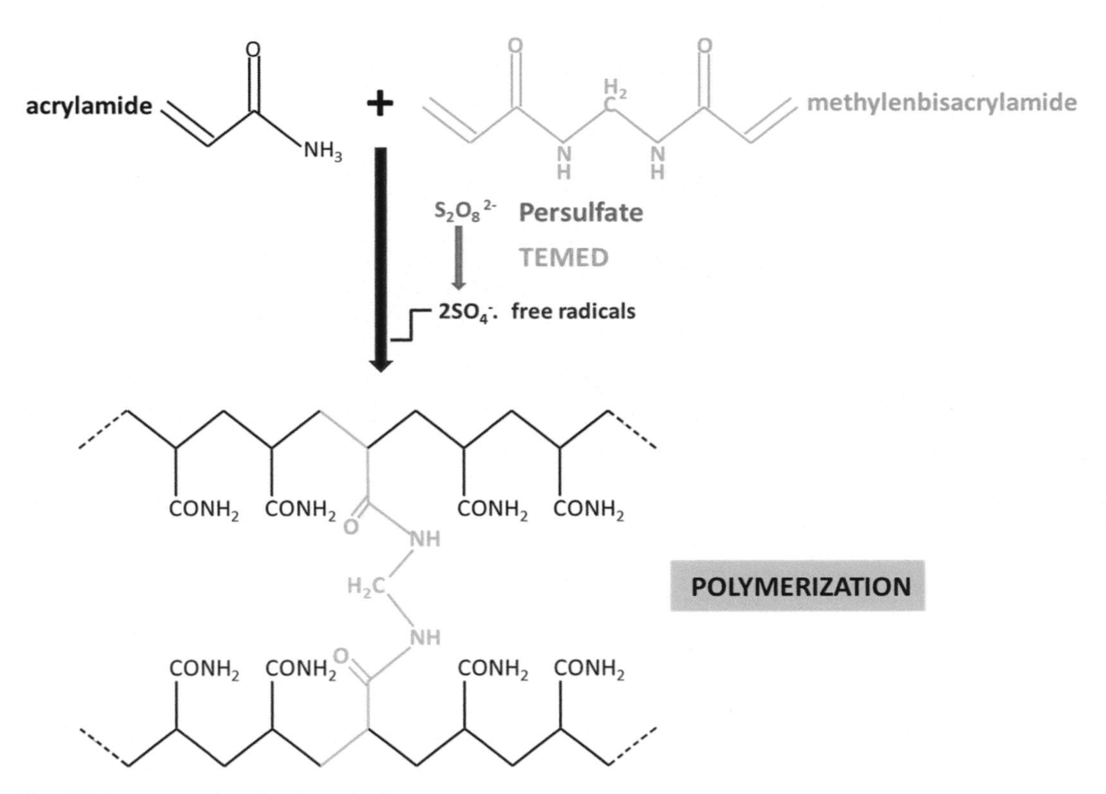

Fig. 5 Polyacrylamide gel polymerization

3. Pour the acrylamide solution into the gap between glass plates using a Pasteur pipette, avoiding the formation of air bubbles.
4. Fit the comb between the glass plates.

After polyacrylamide gel is polymerized it is necessary to make a pre-electrophoresis in 0.5× TBE at 200 V and 4 °C for 2 h to remove traces of APS, urea, and residual polyacrylamide and achieve the application of samples without diffusion (*see* **Note 9**).

A device capable of supplying up to 500 V and 200 mA into a vertical electrophoresis system is necessary.

3.5.2 Electrophoresis

1. Prepare the running buffer (0.5× TBE buffer).
2. Mix each sample with 5 μl of loading buffer with bromophenol blue. Bromophenol blue is a small, colored molecule used to visualize the running of the samples in the electrophoresis.
3. Load each sample onto the gel.
4. Start running at 200 V for 100 min at 4 °C, with an ice block to avoid overheating (*see* **Note 10**).

3.5.3 Determination of Labeling Efficiency

The efficiency of the labeling reaction is controlled through the signal analysis of a series of dilutions of the labeled ds-oligonucleotides, including one control (*see* **Note 11**).

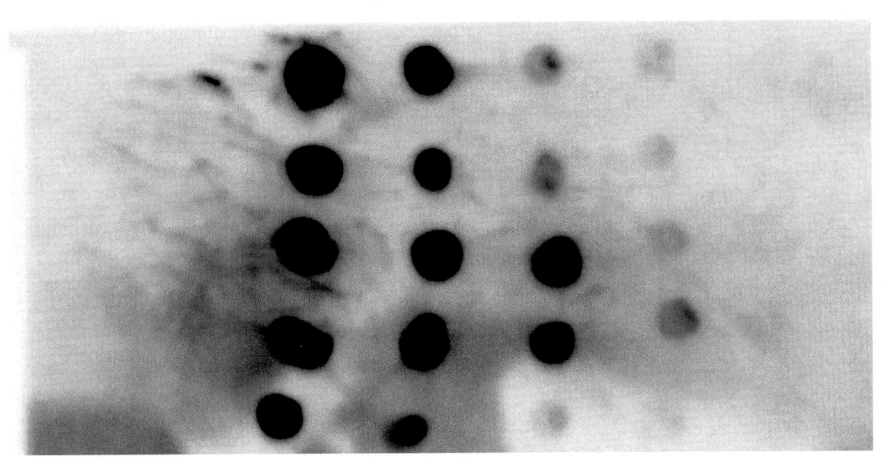

Fig. 6 Labeling efficiency. Analysis of five labeled oligonucleotides (*rows*) and four serial dilutions 1/1, 1/10, 1/100, and 1/1000 and negative control (5 columns, from the left to the right)

1. Prepare four serial dilutions of the labeled ds-oligonucleotides to 1/1, 1/10, 1/100, and 1/1000 in TEN buffer using the same buffer as negative control.
2. Apply 2 μl of each dilution and the negative control in a positively charged nylon membrane and continue with the chemiluminescent detection step. Determine the efficiency of each labeling reaction by comparing each dilution series, considering an optimal labeling whenever 1/1000 dilution can be detected (Fig. 6).

3.6 Blotting and Cross-Linking

1. Once the electrophoresis is finished, carefully remove the glasses from the gel.
2. Equilibrate the nylon membranes for 5 min in transfer buffer (0.5× TBE buffer).
3. Place the equilibrated nylon membrane onto the gel avoiding air bubbles. A Falcon tube can be rolled over the gel and the membrane to eliminate any bubble.
4. Equilibrate the Whatman 3 MM layers for 5 min in transfer buffer (0.5× TBE buffer) and add four layers on one side and other four layers in the other side.
5. Put the resulting sandwich into the electro-transfer system with the nylon membrane toward the positive pole (Fig. 7).
6. Perform the transfer at 4 °C and 400 mA for 30 min.
7. After electro-transfer, place the membrane on a Whatman 3 MM paper pre-equilibrated with SSC.
8. Cross-link at 120 mJ in the cross-linker.

A wet transfer system is used because it is the method of choice in the case of small DNA fragments separated by electrophoresis in polyacrylamide gels. Nucleic acids lesser than 50 pb bind to nylon

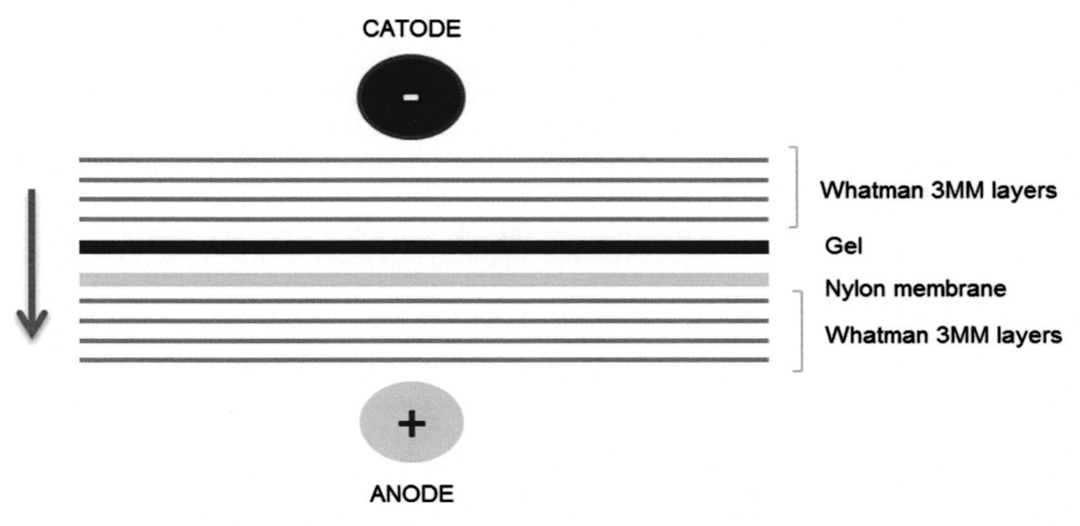

Fig. 7 Orientation of the gel:membrane sandwich in the electro-transfer system [6]

membranes loaded under conditions of low ionic strength of the buffer [8]. Nylon membranes positively charged with amino groups present greater union of nucleic acids, which occasionally brings more background to the hybridization (*see* **Note 3**) [7].

3.7 Chemiluminescent Detection

The detection of this type of labeling is performed with the polyclonal anti-digoxigenin (anti-DIG) antibody conjugated with alkaline phosphatase (AP). The substrate of the enzyme is a chemiluminescent compound, known as CSPD [6]. The enzymatic dephosphorylation of the CSPD by AP produces a phenolate anion that decomposes and delivers light at a wavelength of 477 nm, which is recorded on X-ray films, cameras, or instruments that detect this signal. The detection sensitivity of the DIG labeling depends on the method used to visualize the conjugated antibody. Chemiluminescence is the fastest and more sensitive methodology and the CSPD is one of the most sensitive substrates [7, 9].

1. Once the DNA is transferred and cross-linked to nylon membrane, incubate the membrane for 2 min with the wash solution (1× washing buffer dilution from 10× stock with distilled water).
2. Saturate it with the blocking solution (1× blocking solution dilution from 10× stock with 1× maleic acid buffer) (*see* **Note 12**). The blocking of nonspecific binding sites in the membrane is held at room temperature for 30 min with gentle agitation.
3. Incubate the membrane with antibody anti-digoxigenin-AP (dilution 1:10,000 with blocking solution) at room temperature for 30 min in gentle agitation.
4. In order to eliminate the excess of attached antibody, perform two washes with 1× washing buffer for 15 min each one.

5. Equilibrate the membrane with detection buffer (1× detection buffer dilution from 10× stock with distilled water) for 5 min at room temperature in agitation to adjust pH to 9.5, which is a necessary condition for the CSPD chemiluminescent reaction.
6. Dilute the CSPD 1:100 in 1× detection buffer solution in a total volume of 1 ml and incubate the membrane for 5 min at room temperature. To do this, put the membrane over a clear film with the DNA side face up and disperse the CSPD solution with a pipet over the DNA side of the membrane. Finally wrap the membrane with the film avoiding the presence of air bubbles.
7. Remove excess of substrate, and immediately place the membrane in a second transparent film avoiding the presence of air bubbles (very important). Incubate for 10 min at 37 °C to activate the CSPD chemiluminescent reaction.
8. The generated luminescent compound impresses an auto radiographic high-sensitive film to which is exposed the membrane (Fig. 8) (*see* **Note 13**).

4 Notes

1. Short DNA fragments are cheaper and easily synthesized, and it will probably have less number of nonspecific binding sites, something to consider when the binding specificity is low. Moreover it is possible to get high-resolution levels with short electrophoresis times. The mobility shift on protein binding is bigger than with longer DNA fragments. Small nucleic acids may have structural and electrostatic end effects [10].
2. It is possible to use different protein extracts, nuclear (very useful in analyzing the regulatory elements of gene promoter), cytoplasmic, or whole extracts. Whenever possible, nuclei purification procedures using sucrose pads are preferred in order to eliminate contamination by cytosolic proteins, which more often contain substantial amounts of proteases. Degradation of nuclear proteins by endogenous proteases can be prevented by further addition of protease inhibitors to the buffers [11]. It is important to remember that protein extracts must be kept cold and carefully put them at room temperature just for the strictly necessary time.

   Another problem that can arise when using them is salt excess, which has to be removed. The DNA binding ability of some nuclear proteins has proved to be highly dependent on salt concentration in the reaction mix. It is even useful to evaluate gel shift result depending on KCl concentration in the binding buffer [11].
3. The nylon positively charged transfer membrane binds DNA covalently by UV cross-linking producing strong signal with

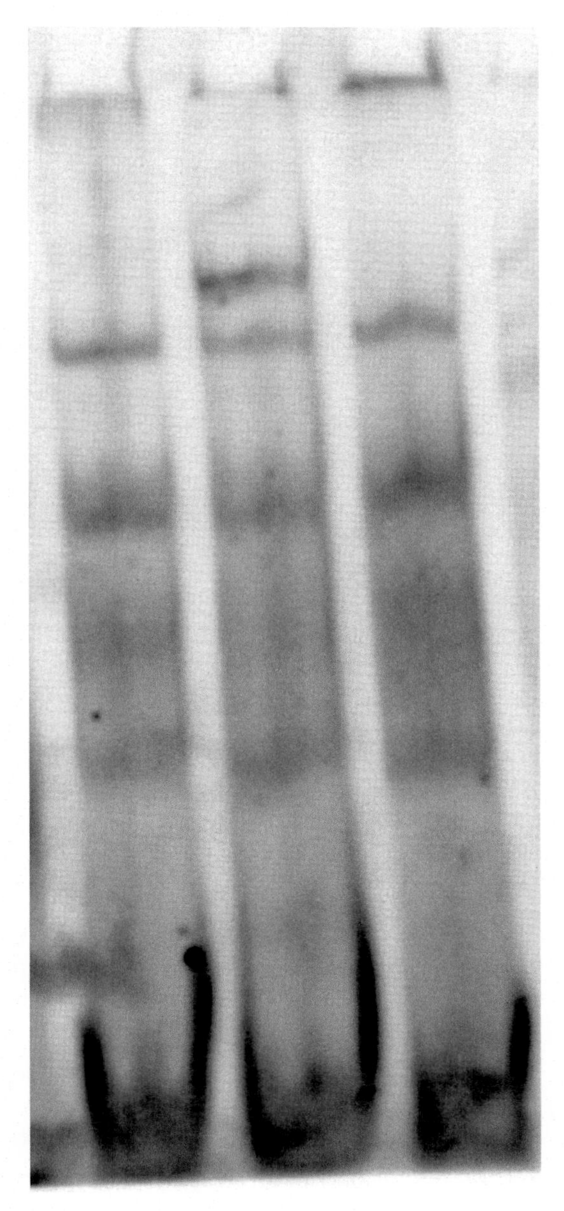

Fig. 8 Example of shift assay. The first lane (*left*), with the labeled ds-oligonucleotide, the second one (*middle*), the labeled ds-oligonucleotide and protein extract, and the third (*right*), with the labeled ds-oligonucleotide, protein extract, and competitor ds-oligonucleotide. There is specific protein or proteins in the nuclear extract that bind to the labeled target DNA what can be inferred because the binding reaction with unlabeled oligonucleotides, where the band corresponding with the DNA-binding protein complexes disappears

smaller quantities of DNA. It is specifically designed to allow numerous reprobings and use in nucleic acid transfers under alkaline conditions.

4. Larger amounts of oligonucleotide can be labeled, but the volume and reaction time have to be increased. It is also better labeling the required oligonucleotides in each experiment.
5. As recommended by the manufacturers, the columns have to be immediately used after this spin, because the resin could dry up fast.
6. The concentration of competitor ds-oligonucleotide has to be empirically determined.
7. Optimal binding temperature and incubation time must be empirically determined.
8. Polyacrylamide concentrations can vary between 4 and 12 %, depending on both size of the labeled DNA fragment and resolution of the DNA:protein complexes, although they are typically used from 4 to 8 %. Lower percentage gels may be needed for larger complexes. For very large complexes, agarose gels may be used, but they have lower resolution power [12].
9. Residual unpolymerized acrylamide could affect the DNA:protein complexes running into the gel. It is recommended to discard the residual acrylamide and use pre-run gels.
10. The given electrophoresis conditions were for two gels in the same running. Gels can be run singly or in pairs, half current for one gel will probably be enough. The optimum migration conditions must be empirically determined for the DNA:protein interaction of interest, using bromophenol blue as reference of the relative migration of free DNA [13].
11. It is possible to use a DIG-labeled control ds-oligonucleotide to compare the spotted series with the unlabeled one after the labeling steps. In first experiments, it is recommendable to determine the labeling efficiency before continuing with the binding reaction.
12. Always prepare fresh blocking solution. The 10× blocking stock is very sensitive to contamination; it is recommended to aliquot it and to work in sterile conditions when possible.
13. The luminescence signal increases in the first few hours, and it remains until 48 h, but it is recommended to expose the membrane to autoradiography films or imaging devices as soon as possible, and take different exposures.

Acknowledgements

This work was supported by the project "Efecto del Ácido Retinóico en la enfermedad alérgica. Estudio transcripcional y su traslación a la clínica," PI13/00564, integrated into the "Plan Estatal de I+D+I 2013–2016" and cofunded by the "ISCIII-Subdirección General de Evaluación y Fomento de la investigación" and the European Regional Development Fund (FEDER).

References

1. Isidoro-Garcia M, Sanz C, Garcia-Solaesa V et al (2011) PTGDR gene in asthma: a functional, genetic, and epigenetic study. Allergy 66:1553–1562
2. Oguma T, Palmer LJ, Birben E et al (2004) Role of prostanoid DP receptor variants in susceptibility to asthma. N Engl J Med 351:1752–1763
3. Revzin A (1989) Gel electrophoresis assays for DNA-protein interactions. Biotechniques 7:346–355
4. Scientific T (2012) http://www.piercenet.com/proteomics/browse.cfm?fldid=7edac33e-3981-4e58-8a57-057a8280c68f
5. Fried M, Crothers DM (1981) Equilibria and kinetics of lac repressor-operator interactions by polyacrylamide gel electrophoresis. Nucleic Acids Res 9:6505–6525
6. Roche Applied Science MA (2006) Instruction manual. DIG gel shift kit, 2nd generation (http://www.protocol-online.org/forums/uploads/monthly_01_2010/post-15212-1263167826.ipb)
7. Russell SA (2001) Molecular cloning. A laboratory manual, vol 2. Cold Spring Harbor Laboratory Press, Cold Spring Harbor, NY, USA
8. Reed KC, Mann DA (1985) Rapid transfer of DNA from agarose gels to nylon membranes. Nucleic Acids Res 13:7207–7221
9. Didenko VV, Hornsby PJ (1996) A quantitative luminescence assay for nonradioactive nucleic acid probes. J Histochem Cytochem 44:657–660
10. Hellman LM, Fried MG (2007) Electrophoretic mobility shift assay (EMSA) for detecting protein-nucleic acid interactions. Nat Protoc 2:1849–1861
11. Gaudreault M, Gingras ME, Lessard M et al (2009) Electrophoretic mobility shift assays for the analysis of DNA-protein interactions. Methods Mol Biol 543:15–35
12. Williams TL, Levy DL (2013) Assaying cooperativity of protein-DNA interactions using agarose gel electrophoresis. Methods Mol Biol 1054:253–265
13. Powell L (2013) Analysis of DNA-protein interactions using PAGE: band-shift assays. Methods Mol Biol 1054:237–251

Chapter 8

SDS-Polyacrylamide Electrophoresis and Western Blotting Applied to the Study of Asthma

Virginia García-Solaesa and Sara Ciria Abad

Abstract

Western blotting is used to analyze proteins after being separated by electrophoresis and subsequently electro-transferred to a membrane. Once immobilized, a specific protein can be identified through its reaction with a labeled antibody or antigen. It is a methodology commonly used in biomedical research such as asthma studies, to assess the pathways of inflammatory mediators involved in the disease.

Here, we describe an example of western blotting to determine the factors involved in asthma. In this chapter, the methodology of western blotting is reviewed, paying attention on potential problems and giving interesting recommendations.

Key words Antibody, Electrophoresis, Horseradish peroxidase, PVDF membranes, Semidry electrophoretic transfer, SDS-polyacrylamide, Western blotting

1 Introduction

The inflammatory alteration involved in asthma can include, among others, the release of proinflammatory mediators from mast cells, the activation of Th2 (T helper type 2) lymphocytes, and the recruitment and degranulation of eosinophils [1]. The study of expression profiles of these proinflammatory mediators has been used to investigate the pathophysiological mechanisms, as well as to search biomarkers of the disease [2–6].

Western blotting or immunoblotting, described by Renart [7] and Towbin [8] in 1979, is a methodology frequently used in the analysis of proteins. In asthma studies, many functional analyses have tried to assess the proinflammatory mediators on different cellular types through this strategy [9–12]. The western blotting can be used to analyze a protein extract, in order to interpret the results of other studies such as gel retardation assays, super-shift assays, or shift-western blotting.

María Isidoro-García (ed.), *Molecular Genetics of Asthma*, Methods in Molecular Biology, vol. 1434,
DOI 10.1007/978-1-4939-3652-6_8, © Springer Science+Business Media New York 2016

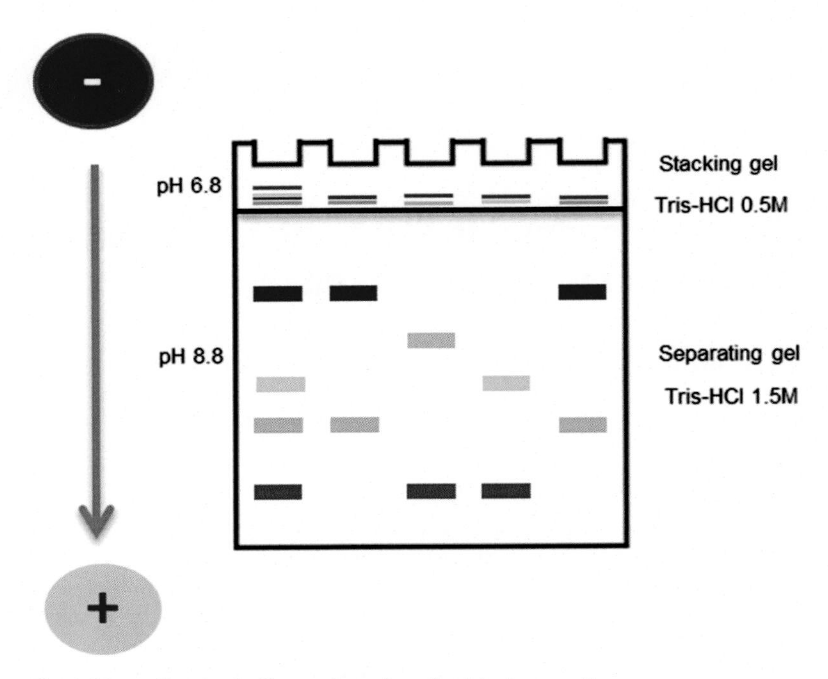

Fig. 1 Discontinuous buffer system described by Laemmli

The electrophoresis of proteins is, nearly always, carried out on denaturing conditions to ensure their dissociation into individual polypeptides and to reduce aggregation. Usually the anionic detergent sodium dodecyl sulfate (SDS) with a reducing agent (DTT or β-mercaptoethanol) and a heat treatment is used to dissociate the proteins before the electrophoresis.

The polypeptides obtained from the denaturing process bind SDS independently of its sequence but proportionally to their molecular weight, becoming negatively charged, in order to migrate toward the anode when a voltage is applied. According to this, the migration in the gel will depend on their size; smaller proteins travel more easily, and faster than the larger ones [13, 14].

Typically the electrophoresis is run in a discontinuous buffer system with two gels, a lower buffer for the separating gel and an upper buffer for the stacking gel. Both contain SDS, but with different ionic strength and pH, as Laemmli described in 1970 [15]. The staking gel concentrates the sample because it does not allow the mobility of the SDS-peptide complexes at its own speed, until they migrate to the separating gel where they are separated according to their size. This allows the separation of relatively large volumes of sample without losing resolution (Fig. 1).

Once separated, the proteins of the gel are transferred to a solid support, such as a nitrocellulose, polyvinyl difluoride (PVDF), or cationic nylon membrane, to be detected using colorimetric, chemiluminescence, or fluorescence techniques. Total protein detection

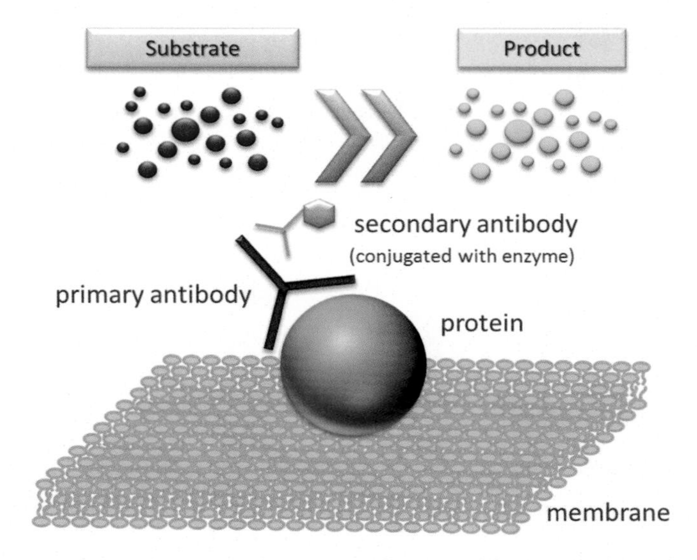

Fig. 2 Western blotting basis. Electro-transferred and immobilized proteins in a membrane can be detected with a labeled specific antibody

allows visualization of the whole content of proteins, while immunodetection methods employ antibodies for the identification of specific proteins, the so-called immunoblotting methodology (Fig. 2).

Some potential difficulties have to be overcome in the setup of this technique, including inefficient transference of proteins, loss of antigenic sites, low sensitivity, high background, or nonquantitative detection [13]. In this chapter, some recommendations are provided to solve or reduce these technical problems.

2 Materials

1. Commercial extract of nuclear proteins from leukocytes of a healthy-individual peripheral blood (AMS Biotechnology). The concentration is 1 mg/ml and the buffer contains HEPES, pH 7.9; glycerol; $MgCl_2$; NaCl; EDTA; and a set of protease inhibitors (*see* **Note 1**).
2. 25–75 ng/μl Positive control in 30 mM Tris–HCl, pH 8.0, 0.2 mM EDTA, 2 mM DTT, 0.2 M NaCl (*see* **Note 2**).

2.1 SDS-Polyacrylamide Gel

1. 40 % (w/v) Acrylamide/Bisacrylamide 29/1 (*see* **Note 3**).
2. 10 % (w/v) Sodium dodecyl sulfate stock solution (*see* **Note 4**).
3. *N,N,N′,N′*-Tetramethylethylenediamine (TEMED) (GE Healthcare).
4. 10 % (w/v) Ammonium persulfate (APS) (Sigma-Aldrich Co.).

2.2 SDS-PAGE Electrophoresis

1. Lower buffer of the separating gel, 1.5 M Tris–HCl, pH 8.8 (*see* **Note 5**).
2. Upper buffer of the stacking gel, 0.5 M Tris–HCl, pH 6.8 (*see* **Note 5**).
3. 5× Tris-glycine running buffer (*see* **Note 6**).
4. 6× Protein sample gel loading buffer (SPLB), 300 mM Tris–HCl, pH 6.8, 15% (w/v) SDS, 0.06% bromophenol blue, 60% (v/v) glycerol, 7.5% β-mercaptoethanol.
5. Distilled water.
6. Protein marker (*PageRuler Prestainer protein ladder plus 10–250 kDa*, Thermo).

2.3 Immunoblotting

1. Membranes of polyvinylidene difluoride, PVDF membranes (Amersham Hybond) (*see* **Note 7**).
2. Whatman 3 MM paper.
3. Electro-blotting device (Trans-Blot SD semidry electrophoretic transfer cell, Bio-Rad) (*see* **Note 8**).
4. 10× Transfer buffer, 386 mM glycine, 66 mM Tris base, 20% MeOH, 0.1% SDS, pH 8.4–8.8 (Promega) (*see* **Note 9**).
5. Pierce Fast Western Blot Kit, Supersignal West Pico Substrate (Thermo Scientific) (*see* **Note 10**).
   - 10× Fast Western Wash Buffer.
   - Fast Western Antibody Diluent.
   - Fast Western Rabbit (or Mouse) Optimized HRP Reagent, Pico.
   - SuperSignal West Pico Luminol/Enhancer Solution.
   - SuperSignal West Stable Peroxide Solution.
6. Antibodies (*see* **Note 11**).
7. High-sensitive autoradiographic films (Kodak®).
8. Kodak® processing chemicals for autoradiography films (Sigma-Aldrich Co.).
9. Darkroom.

3 Methods

3.1 SDS-Polyacrylamide Gel

1. Clean the glass plates with soap and then ethanol. When they are completely clean and dry, properly assemble the glass plates according to the manufacturer's instructions.
2. Mix the components of the separating gel in the order showed in Table 1. The total volume required to completely fill a gel cassette depends on the gel mold and the manufacturer's

Table 1
Components required for preparing the separating gel for SDS-PAGE [13]

Components	10% Gel		8% Gel	
Distilled water	1.95 ml	3.9 ml	2.3 ml	4.6 ml
1.5 M Tris–HCl pH 8.8	1.25 ml	2.5 ml	1.25 ml	2.5 ml
10% SDS	50 μl	100 μl	50 μl	100 μl
10% APS	25 μl	50 μl	25 μl	50 μl
TEMED	2.5 μl	5 μl	2.5 μl	5 μl
30% (29/1) Acryl/bis	1.65 ml	3.3 ml	1.35 ml	2.7 ml
Total volume (ml)	10 (2 gels)	10 (2 gels)	5 (1 gel)	10 (2 gels)

instructions (in this case, employ gels of 0.75 mm thickness that requires 4.2 ml of separating gel). The acrylamide concentration can vary depending on the range of protein separation. The acrylamide solution should be prepared in shake and cold in order to avoid early polymerization. Add the APS and TEMED at the end. The APS is used as a catalyst for the co-polymerization of bisacrylamide and acrylamide gels. The polymerization reaction is caused by the free radicals generated in oxidation-reduction reactions. TEMED participates as a co-catalyst of APS; the polymerization begins as soon as the TEMED is added to the mixture.

3. Pour the separating solution into the gap between glass plates using a Pasteur pipette, avoiding the formation of air bubbles. Leave approximately 2 cm (the length of the comb teeth plus 1 cm) and pour isopropanol using a Pasteur pipette. Wait for 20–30 min for the complete polymerization.
4. During the polymerization of the separating gel, mix the components of the stacking gel in the order that appears in Table 2. The total volume depends on the gel mold and the manufacturer's instructions, but at the same acrylamide concentration.
5. Eliminate the isopropanol overturning the cassette, and pour the acrylamide solution of stacking gel into the gap between glass plates using a Pasteur pipette, avoiding the formation of air bubbles (very important in this point).
6. Fit the comb between glass plates.
7. After 20–30 min the polymerization will be completed.

3.2 SDS-PAGE Electrophoresis

1. Prepare the running buffer (1× Tris-glycine buffer) from 5× stock.
2. Prepare the protein samples in a total volume of 20 μl. Mix 17 μl of the protein sample with 3 μl of 6× SPLB loading buffer (*see* **Note 12**).

Table 2
Components required for preparing the stacking gel for SDS-PAGE [13]

Components	Gel volume (ml)	
	5 (2 Gels)	2.5 (1 Gel)
Distilled water	3.4 ml	1.7 ml
Tris–HCl pH 6.8 0.5 M	1.26 ml	630 μl
10% SDS	50 μl	25 μl
10% APS	50 μl	25 μl
TEMED	5 μl	2.5 μl
30% (29/1) Acryl/bis	830 μl	415 μl

3. Heat the samples at 100 °C for 10 min to get the complete denaturation of the proteins.
4. Remove the comb from the gel. Wash the wells with water in a Pasteur pipette to remove the rest of non-polymerized polyacrylamide.
5. Spin the denatured samples and put them on ice.
6. Load each sample onto the gel.
7. Apply 10 μl of molecular weight marker used as size indicator (Fig. 3).
8. Start running at 60 V and 40 mA for 20 min at 4 °C until the bromophenol blue reaches the bottom of the stacking gel (avoid overheating with an ice piece inside the cuvette). Then increase the voltage at 80–100 V and 40 mA (constant amperage) for approximately 90 min (until the bromophenol blue reaches the bottom of the separating gel).

3.3 Immunoblotting

The Pierce Fast Western Blot Kit, Supersignal West Pico Substrate system, is used for western blot assays that reduce time in different steps of the methodology [16]. It includes secondary antibodies conjugated with horseradish peroxidase (HRP) and luminol substrate for this chemiluminescent assay. This compound is oxidized in the presence of peroxide of hydrogen (H_2O_2) to produce the 3-aminophthalate in excited state, which decays by emitting light recorded on X-ray film, a camera, or an instrument that detect the signal (Fig. 4). The emitted light is captured by a film camera, or more recently by chemiluminescent systems, which take a digital image of the western blotting.

1. Once the electrophoresis is finished, carefully remove the glasses from the gel and equilibrate it in transfer buffer (trim away any stacking gel).

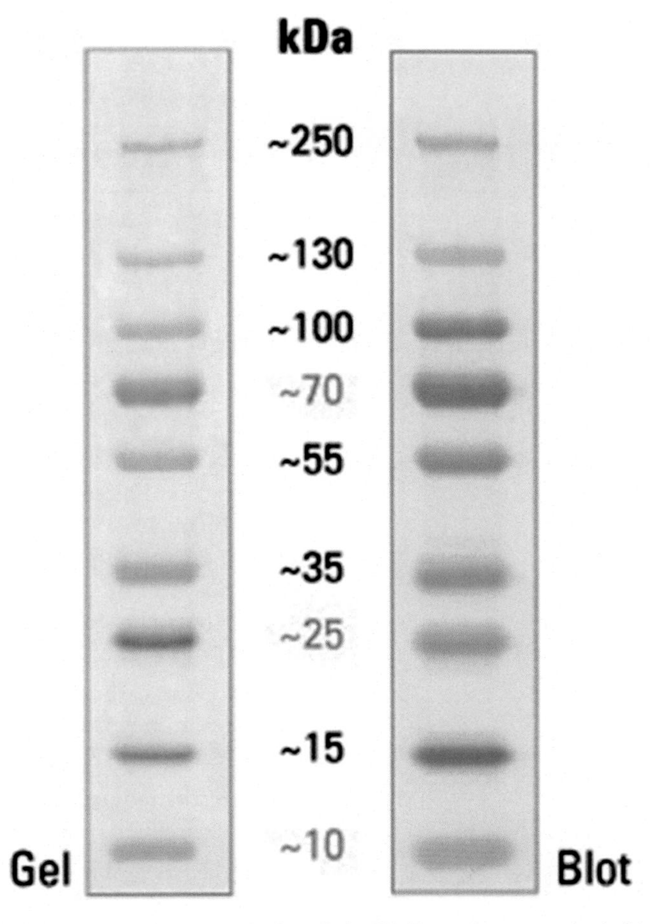

Fig. 3 Molecular weight marker PageRuler™ Prestainer protein ladder plus 10–250 kDa. Size in kilo Daltons (kDa)

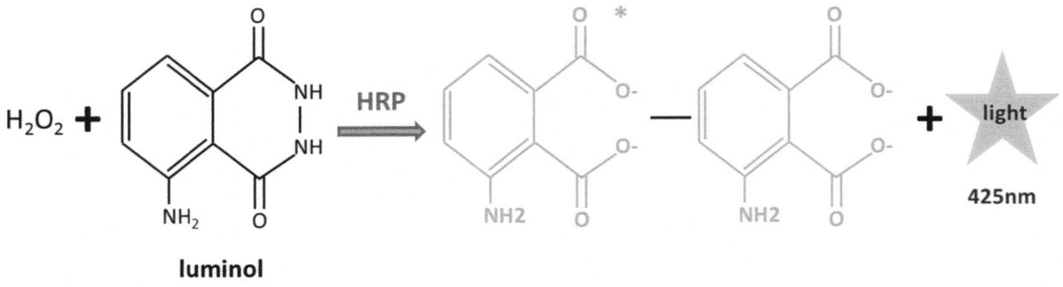

Fig. 4 Oxidation reaction of luminol by HRP

2. Cut the membranes and Whatman papers to gel size, and wet the PVDF membrane with methanol for 10 s.
3. Equilibrate the PVDF membrane and Whatman papers in transfer buffer. The membrane should uniformly change from opaque to semitransparent.

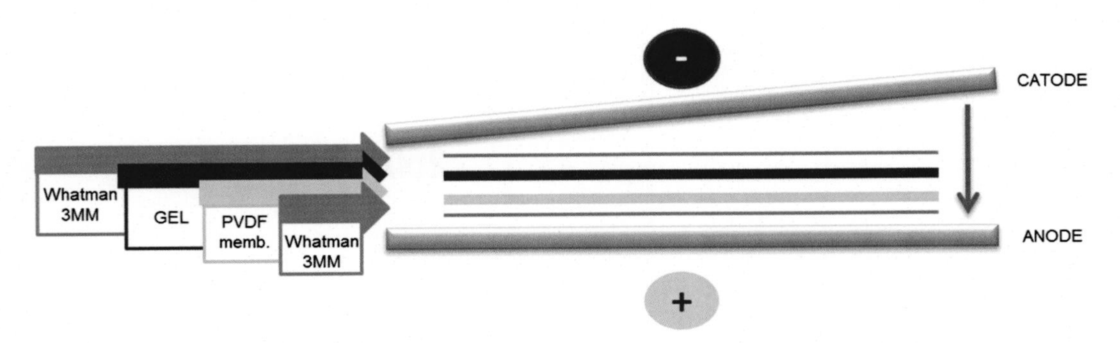

Fig. 5 Scheme of the semidry electrophoretic transfer system. The membrane is placed between the gel and the anode

4. Place a pre-wetted Whatman sheet onto the platinum anode, and roll a pipet or Falcon tube over the surface to roll out air bubbles (*see* **Note 13**). Place the pre-wetted membrane on the Whatman paper, roll out air bubbles, and then place the gel on the membrane. Roll out all air bubbles. Finally place the other pre-wetted Whatman sheet on the top of the gel and roll out air bubbles [17] (Fig. 5).
5. Carefully place the cathode and close the transfer system. Press to engage the latches with the guideposts without disturbing the filter paper stack [17] (Fig. 3).
6. Transfer is performed at 15 V for 40 min (*see* **Note 14**). Once the transfer is finished, withdraw everything and remove the membrane to a clean tray (*see* **Note 15**).
7. Incubate the membrane for 2 min in wash solution (1× wash buffer dilution from 10× stock with distilled water).
8. Incubate the membrane in primary antibody solution (1 μg/ml of the primary antibody/antibodies in blocking solution, Fast Western Antibody Diluent) for 12 h at 4 °C with shaking (*see* **Note 16**).
9. Place the membrane in a clean tray to incubate it at room temperature for 20 min and shaking with secondary antibodies (Fast Western Mouse Optimized HRP Reagent, Pico and/or Fast Western Rabbit Optimized HRP Reagent, Pico) diluted to 1:10 in blocking solution (Fast Western Antibody Diluent) (*see* **Note 17**).
10. Wash the membrane twice in 20 ml of wash solution (1× washing buffer dilution from 10× stock with distilled water) by shaking for 10 min.
11. Incubate the membrane in detection buffer SuperSignal West Pico Working Solution for 5 min at RT (freshly prepared, immediately before to be used; 1:1 dilution of Supersignal West Pico Luminol/Enhacer Solution and Supersignal West Stable Peroxidase Solution).

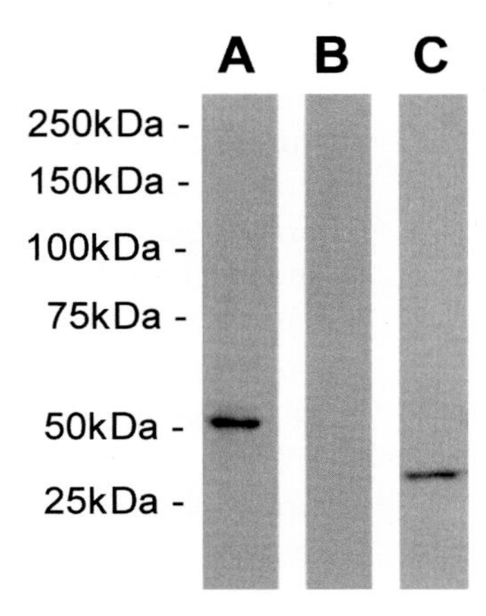

Fig. 6 Western blotting of GATA2 expression. Immunoblotting obtained using GATA binding protein 2 antibody (**a**), negative control (**b**), immunoblotting obtained using anti-beta actin antibody as internal control (**c**)

12. Place the membrane in transparent plastic avoiding the presence of air bubbles (very important) and remove excess of substrate.
13. The generated luminescent compound impresses the high-sensitive autoradiographic film to which the membrane is exposed (*see* **Note 18**).

Analysis of the protein extract by electrophoresis in denaturing conditions and western blotting bring us relevant data, not only in terms of the sample of proteins, but also in the evaluation of own antibodies (Fig. 6).

4 Notes

1. It is possible to use different protein extracts, nuclear, cytoplasmic, or whole extracts. The extraction buffer will be different depending on the protein (acid or basic) to be detected. Always should be done in cold and with protease inhibitors to prevent protein degradation [14]. This strategy can be used to investigate the expression of different proteins, in different tissues (peripheral blood, T-lymphocytes, alveolar macrophages) from asthmatic patients [11]. It is also important to consider the possibility of measuring the concentration of the protein extract, to charge a known amount of proteins in the gel to obtain quantitative or semiquantitative results.
2. It is recommended to use a positive control with the protein of interest (protein recombinant, cells transfected with the protein of interest or tissues or cell lines that express the protein). It is also necessary to include a sample that does not express the protein to detect (negative control; for example from tissue or cell lines that do not express the protein). In expression

studies a load control is completely essential, which is a protein highly conserved and present in similar amounts regardless of the type of sample. Some of the proteins used as loading controls are actin, GAPDH, tubulin, or lamin B1, among others. The choice depends on the protein extract we were going to analyze (whole, cytoplasmic, mitochondrial, and nuclear), the tissue, or the origin and the size range.

3. Most of SDS-polyacrylamide gels are cast with the ratio 1:29 of acrylamide/bisacrylamide, because it resolves differences of 3% in peptide size [13]. It is recommended to have a 29:1 stock solution, ready to use, that is normally stable when stored in dark bottles at room temperature (commercial ones at least 2 years). The gel sieving depends on the ratio of acrylamide and bisacrylamide concentrations, because they form pores with specific size through which SDS-peptide complexes have to pass to be separated.

4. It is recommended to have a 10% SDS stock solution that should be prepared with distilled water, and the same commercial product; do not mix different brands.

5. To prepare lower and upper Tris buffer, the Tris base is diluted in distilled water to obtain 1.5 and 0.5 M concentrations, and then adjust pH with HCl (to 8.8 and 6.8, respectively). Do not use Tris–HCl to prepare buffers, because the salt concentration would be too high and the migration of the peptides through the gel would be abnormal, resulting in bands with diffusion effect [13].

6. It is recommended to have a stock of the running buffer (5×). Dilute 15.1 g of Tris base, 72 g of glycine pH 8.3, and 5 g of SDS with distilled water to a total volume of 1000 ml.

7. Other types of membranes could be used for immunoblotting, nitrocellulose and nylon, which present different properties. To the overall success of a chemiluminescent western blotting the selection of membrane is very important. The nitrocellulose membrane is usually used for western blotting, but is characterized by a small pore size, so it is really used for small proteins. Positively charged nylon membranes are designed to allow for numerous reprobings, but they have some disadvantages; they are difficult to block and can present a high background, especially with more sensitive methods such as chemiluminescence.

   PVDF presents a strong interfacial interaction with proteins (hydrophobic). Before transfer, it is necessary to wet the hydrophobic surface of the membrane with methanol. Proteins bind more tightly to PVDF membranes than to nitrocellulose and are immobilized more efficiently during the detection steps [13].

8. The "semidry" method uses Whatman 3 MM paper saturated with transfer buffer as reservoir. The wet transfer uses vertical electrodes, which are submerged in the transfer buffer, and is suitable for large or difficult-to-transfer proteins, but it required more time. The electrophoretic transfer of proteins seems to be more efficient than capillary transfer. The conditions used for transfer vary according to the design of the specific system, and it is therefore best to follow the manufacturer's instructions at this stage [13].
9. The pH of transfer buffers is 8.3, which is higher than the isoelectric point (pI, IEP) of most of the proteins. Thus, proteins have a net negative charge and migrate toward the positive pole. Methanol added to transfer buffer stabilizes the gel and improves the binding between proteins and PVDF membrane, what somehow prevents the SDS.
10. Other blocking solutions could be used:
    - Powdered milk: 5 % skimmed milk in PBS/TBS. Very economical, with little background, but it deteriorates quickly. It is possible that cover some antigens.
    - Milk/Tween®-20: 5 % of milk in PBS/TBS and 0.1 % Tween®-20. Similar to the previous described, but better to reduce the background.
    - Tween®-20: 0.1 % Tween®-20, 0.02 % sodium azide in PBS/TBS. Economical and it allows staining after the immunodetection although a residual "background" could appear.
    - BSA: 3 % BSA, 0.02 % sodium azide in PBS. Good and relatively economic.
    - Horse serum: 10 % serum of horse, 0.02 % sodium azide in PBS. Without background but it is more expensive and non-compatible with some anti-immunoglobulin antibodies.
    - The blocking component is a nonionic detergent or an inert protein that prevents nonspecific bindings without modifying the interaction of the protein of interest with the membrane. The blocking agent should not alter the recognition of the antibodies to the protein; it neither cross-reacts with the antibodies, nor interferes with the detection method, with the enzymatic reaction or with the reaction product that is going to be quantified to detect the protein of interest.
11. Either polyclonal or monoclonal antibodies can be used, and when possible choose the primary antibodies from the same origin, to share a common secondary antibody in order to cheapen the study. In the case of previously denatured proteins, remember that antibodies recognizing native proteins

obviously cannot be used. Take care with cross-reactions with other proteins or with contaminants for example in the blocking buffer.

In the selection of antibodies we must consider the applications; the manufacturer provides us some indications for its use, in terms of recommended concentration for each application. The concentration of primary and secondary antibodies should be optimized for each experiment, to achieve the greatest possible sensitivity, minimizing the overall background and avoiding nonspecific signals. The results obtained from the western blotting help us to modify the antibody dilution. Usually a high overall background can be reduced by a higher dilution of the secondary antibody while nonspecific signal can be eliminated by higher dilution of the primary antibody or starting with less protein concentration on the electrophoresis.

Antibodies are incubated with the membrane, and diluted in blocking solution to avoid nonspecific binding. In fact, the concentration of the blocking component, as Tween®-20 or other detergent, can be increased to reduce the background.

12. The amount of analyzed protein depends on its relative presence and on the sensitivity and specificity of the primary antibody, as previously discussed. It may be necessary loading several wells with different concentrations of protein and antibody to start working with this method.

13. Bubbles between gel, membrane, and Whatman papers can cause areas of bad or non-transfer on the blot. It is very important to remove it by rolling a pipette or tube.

    Adjust the conditions of transfer to the size of the protein. Smaller proteins will be transferred more quickly and can pass through the membrane. Reduce or eliminate the SDS in the buffer transfer, or use a membrane with a lower pore size to get an optimum retention of the protein. Increase the amount of SDS and reduce the concentration of methanol to facilitate the transfer of large proteins.

14. Do not allow overheating during the transfer. It is recommended to use cold buffer and a cooling system or make the transfer in a cold room. Decrease voltage or time if necessary. It may even be necessary to choose other transfer system. For example, in semidry transfer cells, the electrode plates dissipate a little heat, but they are not recommended for prolonged transfers.

    Incomplete transfer could happen because the current or the time of transfer is insufficient. On the other hand, when the current is too strong, the transfer may also result inefficient because the proteins may migrate through the membrane too fast to be retained; this is interesting to be considered with small proteins. See instructions of the device manufacturer also.

15. To check the efficiency of the transfer, the membranes could be reversibly stained allowing further immunological detection with Ponceau S Staining solution for 5 min at room temperature and agitation. Ponceau S is a negative stain, which binds to the positively charged amino groups of the protein. After that, they need to be washed in distilled water for 2 min and you can see the banding pattern staining in red color corresponding to the sample well transferred. To fully bleed the membranes, they have to be washed with TBS-Tween® (150 mM NaCl, 10 mM Tris pH 7.3; 0.1% Tween®-20). Another way to check the blotting is to use the pre-stained protein marker because it is possible to check how was the transfer.
16. The incubation time may be reduced to 10 min at room temperature; it has to be adjusted depending on each specific antibody/antigen to determine the incubation time [16].
17. To increase the sensitivity, the incubation can be longer than 10 min; however, one or more additional washes will be required to reduce background [16].
18. It is recommended exposing the membrane to autoradiography films as soon as possible, and to take different exposures to obtain the proper balance between signal and background. It is possible to start with 1 min of exposition. When the film is underexposed (not enough time) you could not be able to detect all signals. When the film is overexposed (longer than needed), the signals may be together, not be well separated, or not be distinguished from the background.

    To orientate the photo put a signal, like a piece of adhesive in the upper right corner of the autoradiography film, although the weight marker could be used as reference when it is not in the middle. The emitted light is detected by a CCD camera equipped with appropriate wavelength filters or by a device capable of measuring the fluorescence.

Acknowledgements

This work was supported by the project "Efecto del Ácido Retinóico en la enfermedad alérgica. Estudio transcripcional y su traslación a la clínica," PI13/00564, integrated into the "Plan Estatal de I+D+I 2013–2016" and cofunded by the "ISCIII-Subdirección General de Evaluación y Fomento de la investigación" and the European Regional Development Fund (FEDER).

References

1. Busse WW, Lemanske RF Jr (2001) Asthma. N Engl J Med 344:350–362
2. Pascual M, Roa S, Garcia-Sanchez A et al (2014) Genome-wide expression profiling of B lymphocytes reveals IL4R increase in allergic asthma. J Allergy Clin Immunol 134:972–975
3. Hansel NN, Diette GB (2007) Gene expression profiling in human asthma. Proc Am Thorac Soc 4:32–36
4. Berube JC, Bosse Y (2014) Future clinical implications emerging from recent genome-wide expression studies in asthma. Expert Rev Clin Immunol 10:985–1004
5. Isidoro-Garcia M, Sanz C, Garcia-Solaesa V et al (2011) PTGDR gene in asthma: a functional, genetic, and epigenetic study. Allergy 66:1553–1562
6. Garcia-Solaesa V, Sanz-Lozano C, Padron-Morales J et al (2014) The prostaglandin D2 receptor (PTGDR) gene in asthma and allergic diseases. Allergol Immunopathol 42:64–68
7. Renart J, Reiser J, Stark GR (1979) Transfer of proteins from gels to diazobenzyloxymethyl-paper and detection with antisera: a method for studying antibody specificity and antigen structure. Proc Natl Acad Sci U S A 76:3116–3120
8. Towbin H, Staehelin T, Gordon J (1979) Electrophoretic transfer of proteins from polyacrylamide gels to nitrocellulose sheets: procedure and some applications. Proc Natl Acad Sci U S A 76:4350–4354
9. John AE, Zhu YM, Brightling CE, Pang L, Knox AJ (2009) Human airway smooth muscle cells from asthmatic individuals have CXCL8 hypersecretion due to increased NF-kappa B p65, C/EBP beta, and RNA polymerase II binding to the CXCL8 promoter. J Immunol 183:4682–4692
10. Prefontaine D, Lajoie-Kadoch S, Foley S et al (2009) Increased expression of IL-33 in severe asthma: evidence of expression by airway smooth muscle cells. J Immunol 183: 5094–5103
11. Tomita K, Caramori G, Ito K et al (2012) STAT6 expression in T cells, alveolar macrophages and bronchial biopsies of normal and asthmatic subjects. J Inflamm 9:5
12. Paulissen G, Rocks N, Quesada-Calvo F et al (2006) Expression of ADAMs and their inhibitors in sputum from patients with asthma. Mol Med 12:171–179
13. Sambrook J, Russell DW (2001) Molecular cloning: a laboratory manual, 3rd edn. Cold Spring Harbor Laboratory Press, Cold Spring Harbor, NY
14. Mahmood T, Yang PC (2012) Western blot: technique, theory, and trouble shooting. N Am J Med Sci 4:429–434
15. Laemmli UK (1970) Cleavage of structural proteins during the assembly of the head of bacteriophage T4. Nature 227:680–685
16. Fisher T Instruction Manual. Pierce fast western blot kit, supersignal west pico substrate
17. Bio-Rad Trans-Blot® SD Semi-Dry Electrophoretic Transfer Cell Instruction Manual. Catalog Number 170-3940

Chapter 9

Chromatin Immunoprecipitation: Application to the Study of Asthma

Asunción García-Sánchez and Fernándo Marqués-García

Abstract

Chromatin immunoprecipitation (ChIP) is a technique for studying interactions between proteins and DNA in living cells. A protein of interest is selectively immunoprecipitated from a chromatin preparation, to analyze the DNA sequences involved. ChIP can be used to determine whether a transcription factor interacts with a candidate target gene and to map the localization of histones with posttranslational modifications on the genome.

The protein-DNA interactions are captured in vivo by chemical cross-linking. Cell lysis, DNA fragmentation, and immunoaffinity purification of the protein of interest allow to co-purify DNA fragments that are associated with that protein. The enriched protein-DNA population is ready to be quantified by PCR to detect precipitated DNA fragments. The combination of ChIP with DNA microarray analysis (ChIP-on-chip) and high-throughput sequencing (ChIP-seq) has enabled to obtain profiles of transcription factor occupancy sites and histone modifications throughout the genome.

Key words Antibody, ChIP-on-chip, ChIP-PCR, ChIP-seq, Chromatin, Cross-link, DNA, DNA-protein, Gene expression, Immunoprecipitation

1 Introduction

Chromatin immunoprecipitation is a powerful technique that allows us to analyze in vivo protein-DNA interactions [1]. In eukaryotes, DNA is found in complexes with proteins and RNA. These complexes correspond to the chromatin in which the DNA is packaged around the histone proteins, forming the nucleosomes. Histones play an important role in the regulation of gene expression.

A large number of residues, found on the histones, are modified by processes such as acetylation, methylation among others. These histone modifications disrupt the contact between nucleosomes to "unravel" chromatin and recruit non-histone proteins [2]. ChIP analysis permits to determine the specific location of proteins such as transcription factors or histone and its modifications. The identification of these protein targets provides valuable insights

María Isidoro-García (ed.), *Molecular Genetics of Asthma*, Methods in Molecular Biology, vol. 1434,
DOI 10.1007/978-1-4939-3652-6_9, © Springer Science+Business Media New York 2016

into how these proteins (and their modifications) work in their normal chromatin context. The chromatin immunoprecipitation reveals the protein-DNA interactions in a native chromatin environment [3].

The first chromatin immunoprecipitation assay was developed by Gilmour and Lis [4–6] to monitoring the association of RNA polymerase II with genes in *Escherichia coli* and *Drosophila*. The use of formaldehyde in the ChIP method was developed by Solomon et al. [7] to study the association of histone H4 and RNA polymerase II with the *Drosophila* Hsp70 genes. The PCR, as a detection method, was firstly used by Hecht et al. [8], in their studies of SIR protein interaction in *Saccharomyces cerevisiae*. The adaptation of ChIP to use in mammalian cells was done in 1997 [9–11] and, 1 year later, Hebbes et al. reported the use of antibodies against histone modifications. An antibody recognizing *N*-acetyl-lysine was used to immunoprecipitate nucleosomes containing acetylated histones from chicken erythrocyte nuclei. The immunoprecipitated chromatin was probed with β-globin and ovalbumin DNA sequences. A specific enrichment of the β-globin locus demonstrated a link between histone acetylation and an active transcriptional state in vivo [12].

The use of formaldehyde as a cross-linking reagent is the most extended method for carrying out ChIP experiments, and it is known as X-ChIP. Proteins are reversibly cross-linked to DNA with formaldehyde (which is heat-reversible) to covalently attach proteins to their target DNA sequences. Thus, the DNA-protein complexes are maintained during all ChIP procedure [13]. Formaldehyde is the cross-linker agent suitable for proteins, which directly bind DNA because of the short cross-links arm of formaldehyde. To avoid this restriction, several long-range bifunctional cross-linkers have been successfully used, in combination with formaldehyde to detect proteins that indirectly associate with DNA [14]. Then, chromatin is typically fragmented by sonication creating a smear of DNA suitable for the analysis of the chromatin associate factor.

Other variant of ChIP is performed in the absence of cross-linking, the native ChIP or N-ChIP. Native chromatin is used as the substrate and the fragmentation of chromatin is typically performed by micrococcal nuclease digestion. This ChIP variant is limited to analyze proteins that are stably associated with chromatin like histones and their modifications [15, 16], while X-ChIP is suitable for transcription factors or any other weakly binding chromatin proteins.

The major advantage of N-ChIP is better antibody specificity and consequently a better recovery of chromatin. The chromatin is cleared by sedimentation, removing the cellular debris and protein-DNA complexes that are immunoprecipitated by using antibodies specific for the protein of interest. The complexes are washed

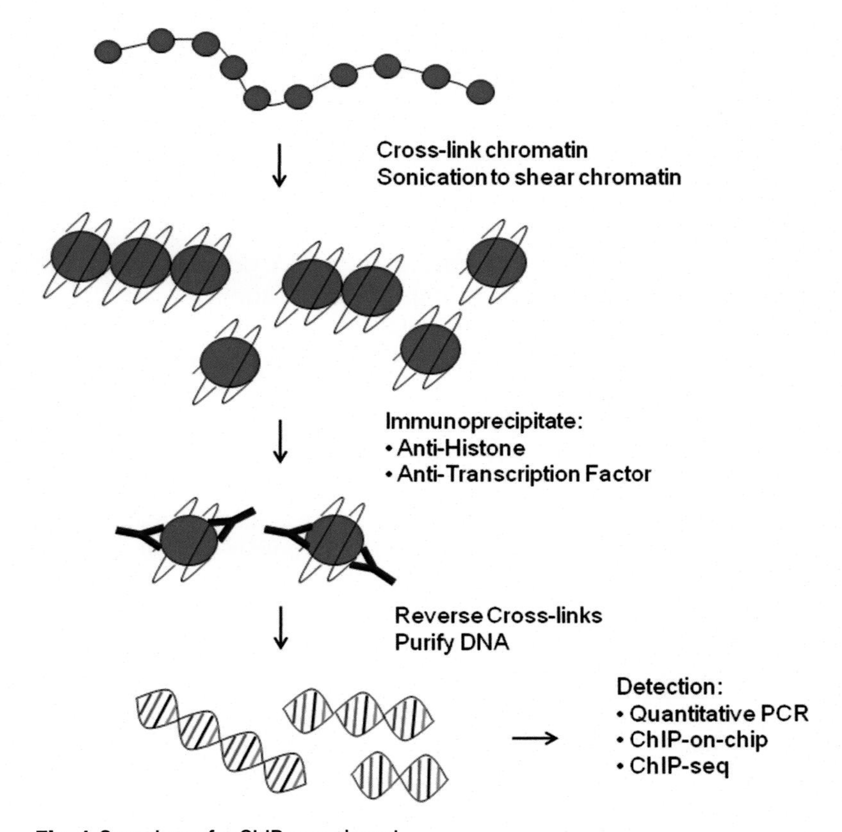

Fig. 1 Overview of a ChIP experiment

under stringent conditions to ensure removal of unspecifically bound chromatin. Then, the cross-link is reversed by heating the sample. The proteins are digested with proteinase K and the immunoprecipitated DNA is purified. The DNA sequences associated with the pulled down protein are identified by polymerase chain reaction (PCR), quantitative PCR (qPCR) (Fig. 1). ChIP has also been applied for genome-wide analysis by combining with microarray technology (ChIP-on-chip) [17, 18] or new-generation DNA-sequencing technology (ChIP-Sequencing) [19–21].

The typical ChIP protocol requires large cell numbers, involves many steps, and is time consuming (usually take 4–5 days), so throughout these decades many modifications have been introduced in order to make it faster, avoiding possible technical errors [12, 15, 16, 22–31]. Many companies have developed different ChIP kits with the purpose of making the technique friendlier. At present, new ChIP modifications could be achieved with only 100–1000 cells and completed in 1 day.

The ChIP-on-chip method combines chromatin immunoprecipitation and DNA microarrays to identify protein-DNA interactions in vivo. The enriched DNA population is eluted after reversal cross-linking and it is labeled with different fluorochromes for the sample and input (usually Cy5 and Cy3). The two populations are mixed and hybridized onto a microarray containing oligonucleotide probes covering whole genome, or probes tiling a region of interest (the promoter region). The binding of the immunoprecipitated transcription factor to a specific region is calculated when the intensity of the ChIP DNA is significantly higher than the input DNA on the array. Computational and bioinformatic approaches are applied to normalize the enriched sample and reference [32].

ChIP-seq consists of the chromatin immunoprecipitation followed by high-throughput sequencing. The immunoprecipitated DNA is reversely cross-linked, fragmented, and analyzed by massive parallel DNA sequencing. This method was first published in 2007 [33–36]. The main advantage of this ChIP is its higher resolution, fewer artifacts, and a great coverage. Here we focus on the typical protocol of chromatin immunoprecipitation and the preparation of the library to the subsequent ChIP-seq procedure.

2 Materials

2.1 Immunoprecipitation

1. Culture cells (50–100 millions per experiment).
2. 37% Formaldehyde.
3. 2.5 M Glycine (Mm 75.07 g/mol): Prepare a stock combining 9.38 g in 50 ml of distilled H_2O. Vortex to completely dissolve the Glycine before use. Sterilize by autoclave. Store at room temperature.
4. 1× Phosphate-Buffered Saline (PBS).
5. 0.1 M Phenylmethylsulfonyl fluoride (PMSF) in ethanol.
6. Protease inhibitor Cocktail, i.e., complete, EDTA-free Protease Inhibitor Cocktail (Roche Applied Science).
7. 10% Triton X-100 detergent solution in water.
8. 10% Nonyl phenoxypolyethoxylethanol (NP40) detergent solution in water.
9. 0.5 M EDTA (Ethylene diamine tetraacetic acid) (Mm 292.24 g/mol): Add 4.38 g to 30 ml of distilled H_2O. Add 10 N NaOH to adjust the solution at pH 8 required to completely dissolve the EDTA. Sterilize by autoclave. Store at room temperature.
10. 100 mM EGTA (Ethylene glycol tetraacetic acid) (Mm 380.4 g/mol): Add 3.8 g to 20 ml of distilled water and bring to pH 11 with NaOH. Adjust pH 7 with HCl and add H_2O to a final volume of 100 ml. Sterilize by autoclave. Store at 4 °C.
11. 1.5 M Tris–HCl pH 8 (Mm 121.4 g/mol): Dissolve 12.1 g in distilled H_2O. Adjust pH to 8.1 with HCl and adjust the

volume to 100 ml. Sterilize by autoclave. Store at room temperature.

12. 1 M Tris–HCl pH 6.5 (Mm 121.4 g/mol): Dissolve 12.1 g in distilled H_2O. Adjust pH to 6.5 with HCl and adjust the volumen to 100 ml. Sterilize by autoclave. Store at room temperature.
13. 5 M NaCl (Mm 58.44 g/mol): Prepare a stock by combining 8.76 g in 30 ml of H_2O. Vortex to completely dissolve it before use. Sterilize by autoclave. Store at room temperature.
14. 10% Sodium Deoxycholate (Mm 414.55 g/mol): Add 5 g into 50 ml of distilled H_2O. Protect from light.
15. 10% Sodium dodecyl sulfate (SDS) (Mm 288.37): Dissolve 10 g of SDS into 80 ml of distilled H_2O and complete with H_2O to a final volume of 100 ml. Store at room temperature.
16. 10 M LiCl (Mm 42.40 g/mol): Dissolve 42.4 g in 10 ml of distilled H_2O. Sterilize by autoclave. Store at room temperature.
17. Protein A/G-PLUS agarose beads pre-blocked with BSA (Santa Cruz).
18. 5 mg/ml Glycogen.
19. Tris-EDTA buffer solution: 10 mM Tris–HCl pH 8, 1 mM EDTA.
20. QIAquick PCR purification kit (Qiagen).
21. 10 mg/ml Proteinase K.
22. 100 mg/ml RNAse A.
23. Red Safe Nucleic Acid Staining Solution (iNtRON, Korea).
24. 100 mg/ml BSA.
25. 10× Stock solution of protease inhibitor cocktail: Complete, EDTA-free MINI Protease Inhibitor Cocktail (Roche Applied Science) (ten mini tablets in 10 ml of water).
26. Nuclear Lysis Buffer: 0.25% Triton X-100, 0.5% NP40, 10 mM EDTA, 0.5 mM EGTA, 10 mM Tris–HCl pH 8. (Add PMSF and protease inhibitor cocktail just before use).
27. Buffer 3: 0.2 M NaCl, 1 mM EDTA, 0.5 mM EGTA, 10 mM Tris–HCl pH 8. Add PMSF and protease inhibitor cocktail just before use.
28. SDS Sonication Buffer: 10 mM EDTA, 50 Mm Tris–HCl pH 8, 1% SDS.
29. ChIP Dilution Buffer: 1.1% Triton X-100, 0.01% SDS, 167 mM NaCl, 16.7 mM Tris–HCl pH 8, 1.2 mM EDTA.
30. Buffer 4: 1 mM EDTA, 0.5 mM EGTA, 10 mM Tris–HCl pH8. Add PMSF and protease inhibitor cocktail just before use.

31. 10× RIPA Buffer: 10 % Triton X-100, 1 % Sodium Deoxycholate, 1.4 M NaCl, 1 % SDS.
32. Elution Buffer: 1 % SDS, 0.1 M $NaHCO_3$.
33. Low Salt Immune Complex Wash Buffer: 0.1 % SDS, 1 % Triton X-100, 2 mM EDTA, 20 mM Tris–HCl pH 8, 150 mM NaCl.
34. High Salt Immune Complex Wash Buffer: 0.1 % SDS, 1 % Triton X-100, 2 mM EDTA, 20 mM Tris–HCl pH 8, 500 mM NaCl.
35. LiCl Buffer: 0.25 M LiCl, 1 % NP-40, 1 % Deoxycholate, 1 mM EDTA, 10 mM Tris–HCl pH 8.

2.2 ChIP-seq Preparation Library Reagents

1. T4 DNA ligase buffer with 10 mM ATP.
2. dNTPs mix.
3. T4 DNA polymerase.
4. Klenow DNA polymerase.
5. T4 PNK.
6. Klenow buffer.
7. dATP.
8. Klenow fragment (3′–5′ exo minus).
9. DNA ligase buffer.
10. Adapter oligo mix.
11. DNA ligase.
12. Ultra pure water.
13. Phusion polymerase (New England Biolabs).
14. 5× Phusion buffer (New England Biolabs).
15. dNTP mix.
16. PCR primer 1.1 (Illumina).
17. PCR primer 2.1 (Illumina).
18. Quant-iT dsDNA HS Assay Kit, 0.2–100 ng (Life Technologies).
19. MinElute PCR Purification kit (Qiagen).

2.3 Equipment

1. Culture dishes or Flasks.
2. Centrifuge.
3. Rotating wheel/platform.
4. Sonicator (i.e., Bioruptor, Diagenode).
5. Electrophoresis system.
6. Thermomixer (Eppendorf).
7. Qubit Fluorometer (Life Technologies) or 2100 Bioanalyzer (Agilent Technologies).
8. Timer.

9. Variable volume (5–1000 μl) pipettors and tips.
10. Low binding microfuge tubes, 1.5 ml.
11. Thermal cycler.
12. PCR tubes, 0.2 ml.

3 Methods

3.1 Immunoprecipitation

3.1.1 In Vivo Cross-Linking and Lysis

Stimulate or treat, if necessary, the mammalian cells in culture. The day before the cross-linking, seed the cells to obtain the correct cell density. This should be 8×10^5 cells/ml for suspension cells or 5–6 culture dishes (15 cm) with cells at 80–90% confluence. Optimization is required depending on the cell line.

The aim of cross-linking is to fix the antigen of interest to its chromatin-binding site. The links formed using formaldehyde are reversible and DNA fragments are easily purified.

1. Add 37% formaldehyde to the growth media to cross-link. Final concentration of formaldehyde is 1% (*see* **Note 1**).
2. Incubate at room temperature for 10 min with rotation gently (*see* **Note 2**).
3. Add Glycine to a final concentration of 0.125 M to the media, and incubate 5 min at room temperature with gentle rotation (*see* **Note 3**).
4. For suspension cells, centrifuge the cells at 370×*g* at 4 °C for 5 min and wash the pellet twice with 6–8 ml of ice-cold 1× PBS. For adherent cells, remove the media and wash cells twice with 6–8 ml of ice-cold 1× PBS, completely remove wash for culture dish each time.
5. Add 5 ml of ice-cold PBS with protease inhibitors (one EDTA-free protease inhibitor complete tablet into 50 ml of ice-cold 1× PBS) to each dish or tube cells. Scrape or resuspend the cells into the cold buffer. Combine cells from different plates into one tube. Centrifuge at 370×*g*, 4 °C for 5 min.
6. Remove completely the PBS (*see* **Note 4**).
7. Resuspend 50×10^7 cells in 2 ml Nuclear Lysis Buffer containing 1× protease inhibitor cocktail. Incubate on ice for 10 min, mix by inverting every 3–5 min, and centrifuge at 2000×*g* at 4 °C for 5 min. Discard the supernatant.
8. Resuspend the nuclei in 2 ml of Buffer 3 containing 1× protease inhibitor cocktail. Incubate on ice for 10 min, mix by inverting every 3–5 min, and centrifuge at 2000×*g* at 4 °C for 5 min.
9. Resuspend the nuclei in 2 ml SDS Sonication Buffer, containing 1× protease inhibitor cocktail. Incubate the cell lysate on ice until sonication.

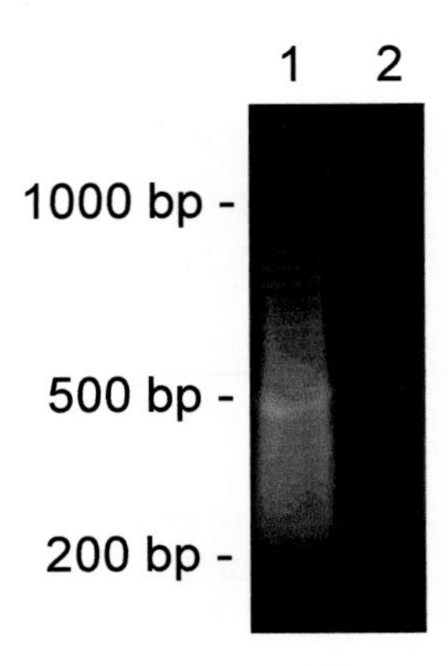

Fig. 2 DNA sonication. Sheared and unsheared chromatin from formaldehyde-cross-linked HEL cells was sonicated using Bioruptor instrument. Perform two run of ten cycles: 30 s "ON" 30 s "OFF" with SDS Sonication Buffer. 10 μl of sheared (*lane 1*) and unsheared (*lane 2*) chromatin was then electrophoresed through a 1 % gel stained with Red Safe. *Lane 1* shows that the majority of the DNA has been sheared between 200 and 1000 bp

3.1.2 Sonication to Shear DNA

Optimal conditions need to be determined to shear the cross-linked DNA to 200–1000 base pairs in length (Fig. 2). The time required depends on the cell line, the volume of cell lysis buffer, cell density, sonication buffer (with/without 1 % SDS), and equipment. Follow the manufacturer's guidelines for specific instructions of your instrument (*see* **Note 5**).

1. Remove 10 μl of cell lysate prior starting the sonication for agarose gel analysis of unsheared DNA and reserve it.
2. Sonicate the cell lysate on wet ice following the recommendations of your instrument (*see* **Note 6**).
3. Remove 10 μl of cell lysate in each sonication run for agarose gel analysis of the sheared DNA.
4. Centrifuge all the aliquots (unsheared and sheared) at 8000 × *g* in microfuge at 4 °C for 5 min to remove cellular debris and keep the supernatant.
5. Add 1.5 μl of 5 M NaCl and incubate at 65 °C for 4–5 h.
6. Add 1 μl of RNAse A (10 mg/ml) and incubate at 37 °C for 30 min.
7. Add 1 μl Proteinase K (10 mg/ml) and incubate at 62 °C for 2 h.
8. Load 10 μl in a 1 % agarose gel and stain with Red Safe. Include a 100 bp DNA marker.
9. Observe which of the shearing conditions gives a smear of DNA in the range of 200–1000 bp, *see* Fig. 2 for an example (*see* **Note 7**).

10. If the resulting DNA smears are not in the desired size range, repeat optimizing the shearing conditions.
11. Centrifuge the sheared sample at 8000 × *g* in microfuge at 4 °C for 5 min to remove cellular debris and keep the supernatant. Mix all supernatants if any. Sheared cross-linked chromatin can be stored at −80 °C for up to 2–3 months.

3.1.3 Chromatin Preclearing and Immunoprecipitation (IP)

1. Sheared chromatin should be diluted ten times with ChIP Dilution Buffer (*see* **Note 8**). Prepare enough ChIP Dilution Buffer containing 1× Protease Inhibitor Cocktail for the number of desired immunoprecipitations and store it on ice. The final volume of each immunoprecipitation should be 1000 ml. It is recommended to include a negative control IgG of the same species as the antibody of interest. Prepare 1.5 ml low binding microtubes, each containing 100 μl of sheared chromatin for the number of desired immunoprecipitations and put on ice.
2. Add 900 μl of ChIP Dilution Buffer containing 1× Protease Inhibitor Cocktail (*see* **Note 9**).
3. To preclear the chromatin, add 30 μl of protein A/G agarose beads to 2 ml of chromatin on ice and incubate at 4 °C for 1 h in a wheel rotator (*see* **Notes 10–12**).
4. Pellet the agarose beads by centrifugation at 3000 × *g* for 1 min and remove supernatant to a new low binding tube.
5. Remove 10 μl (1 %) of the supernatant as “Input” and save at 4 °C (*see* **Note 13**).
6. Remove 1 ml supernatant (~25 × 10^6 cells) to a new 1.5 ml low binding tube avoiding take agarose beads for each IP and add the specific antibody and the IgG antibody control (*see* **Notes 14–16**).
7. Incubate for 1 h to overnight at 4 °C with rotation. Typically the antibody incubation is overnight.

3.1.4 Washes and Elution

1. Add 50 μl Protein A/G-PLUS Agarose Beads to each IP on ice and incubate at 4 °C for 1 h with rotation (*see* **Note 11**).
2. Pellet the Protein agarose beads by centrifuge at 3000 × *g* at 4 °C for 1 min.
3. Carefully remove the supernatant completely avoiding dragging the agarose beads.
4. Wash the Protein agarose beads with 1 ml of Low Salt Buffer at 4 °C for 10 min with rotation.
5. Centrifuge the samples at 3000 × *g* for 1 min at 4 °C.
6. Carefully remove supernatant preventing to drag agarose beads.
7. Wash the Protein agarose beads with 1 ml of High Salt Buffer at 4 °C for 10 min with rotation.

8. Centrifuge the samples at 3000 × *g* for 1 min at 4 °C.
9. Carefully remove supernatant avoiding dragging agarose beads.
10. Wash the protein agarose beads with 1 ml of LiCl Buffer at 4 °C for 10 min with rotation.
11. Centrifuge the samples at 3000 × *g* for 1 min at 4 °C.
12. Carefully remove supernatant avoiding dragging agarose beads.
13. Wash the Protein agarose beads with 1 ml of TE Buffer at 4 °C for 10 min with rotation.
14. Centrifuge the samples at 3000 × *g* for 1 min at 4 °C.
15. Carefully remove supernatant avoiding dragging agarose beads.
16. Resuspend the beads and input samples in 100 μl Elution Buffer.
17. Mix in a Vortex at maximum speed for 20 s and incubate on a thermo mixer at 37 °C and maximum speed for 30 min.
18. Pellet the Protein Agarose beads by centrifuge at 3000 × *g* at 4 °C for 2 min.
19. Collect the supernatant to a new low binding tube.
20. Repeat **steps 16–19** and put together the two supernatants to a final volume of 200 μl.
21. Add 200 μl of Elution Buffer to the 10 μl of Input reserved before.
22. To reverse cross-link, add 12 μl of 5 M NaCl to the samples and input and incubate at 65 °C for 4–5 h to overnight.
23. Add 1 μl of RNase A and incubate at 37 °C for 30 min.
24. Add 4 μl of 0.5 M EDTA, 8 μl of 1 M Tris–HCl, and 1 μl of Proteinase K to each tube and incubate at 45 °C for 1–2 h.

3.1.5 DNA Purification

1. Centrifuge the samples at full speed for 5 min at room temperature.
2. Perform the purification of DNA by spin columns (e.g., QIAquick PCR Purification Kit) or standard phenol: chloroform extraction.
3. Resuspend DNA in two aliquots of 15 μl warmed (~55 °C) Qiagen Elution Buffer; allow entering the column, spin, and repeat.
4. Measure DNA concentration by using a Qubit fluorometer.
5. DNA is ready for end-point PCR, real-time PCR, or Illumina library construction in case to follow with ChIP-seq (*see* **Note 17**).

3.1.6 ChIP-PCR

Real-time PCR is usually the preferred method for analyzing specific DNA fragments. The enrichment of a target depends on several factors: the accessibility of the antigen, the affinity of the antibody, and the precise conditions of the immunoprecipitation. Because of that, the level of enrichment is presented as ratio of precipitated sequence over input. It is named as "percent input" values, calculated by using real-time PCR to quantify the amount of the DNA fragment of interest added to the ChIP reaction, with respect to the amount of the DNA fragment found in the final immunoprecipitate.

It is important designing PCR primers to different regions in the gene of interest. As negative control, primers designed over other genomic DNA fragments that do not precipitate with the specific antibody should be used. A parallel ChIP experiment should be performed with a positive control antibody to ensure no technical failures occur. Typically, the positive antibody controls are anti-RNA Polymerase and Histone H3. The control PCR primers are the human GAPDH gene and human RPL30 gene, respectively.

3.2 ChIP-seq

The immunoprecipitated DNA fragments can be sequenced by high-throughput technology. A library with the ChIP-DNA fragments should be prepared to perform the massively parallel sequencing. First, DNA fragments should be repaired using T4 DNA polymerase, Klenow polymerase and T4 polynucleotide kinase (PNK), which convert overhangs into phosphorylated blunt ends. A control library, e.g., non-immunoprecipitated fragmented DNA (Input sample) or immunoprecipitated DNA sample with an unspecific IgG antibody is also prepared to detect differential enrichment.

3.2.1 DNA Repair

The 3′–5′ exonuclease activity of T4 DNA polymerase, Klenow, and T4 PNK removes 3′ overhangs and the polymerase activity fills in the 5′ overhangs.

1. First dilute the Klenow DNA polymerase 1:5 with water to reach the final Klenow concentration of 1 U/μl.
2. Prepare the following reaction mix in a 1.5 ml low-bind DNA microtube:
   (a) 30 μl purified ChIP-enriched DNA.
   (b) 10 μl water.
   (c) 5 μl 10× T4 ligase buffer with 10 mM ATP.
   (d) 2 μl 10 mM dNTPs mix.
   (e) 1 μl T4 ligase.
   (f) 1 μl Klenow.
   (g) 1 μl PNK.

The total volume should be 50 μl.

3. Incubate in a Thermomixer or a thermo cycler at 20 °C for 45 min.
4. Purify DNA using the MinElute PCR Purification kit and elute with 41 μl of Elution Buffer (Qiagen).

3.2.2 Addition of "A" Bases to the 3′-end of Fragments

This protocol adds an "A" base to the 5′ phosphorylated blunt ends using the polymerase activity of Klenow 3′–5′ exo minus.

1. Prepare the following reaction mix in a 1.5 ml low-bind DNA microtube:
   (a) 41 μl blunted DNA.
   (b) 1 μl of 10 mM dATP.
   (c) 5 μl NEB Buffer 2.
   (d) 3 μl Klenow exo (3′–5′ exo minus).
2. Incubate at 37 °C for 45 min in a thermomixer or thermal cycler.
3. Purify DNA using the MinElute PCR purification Kit and elute with 16 μl Elution Buffer.

3.2.3 Ligate Adapter to DNA Fragments

At this step the adapters are ligated to ends of the DNA fragments, preparing them to the hybridization.

1. Dilute the adapter oligo mix 1:10 with water to adjust to the smaller quantity of DNA (*see* **Note 18**).
2. Prepare the following reaction mix in a 1.5 ml low-bind DNA microtube:
   (a) 16 μl DNA sample.
   (b) 2 μl 10× Ligase Buffer.
   (c) 1 μl Diluted Adapter Oligo Mix.
   (d) 1 μl T4 ligase.

   Total volume should be 20 μl.
3. Incubate at room temperature for 6 h and purify DNA (*see* **Note 19**).
4. Purify DNA using the MinElute PCR Purification Kit and elute with 10–15 μl of Elution Buffer.

3.2.4 PCR with Illumina Primers

1. Prepare the following PCR reaction mix:
   (a) 10–15 μl DNA (all).
   (b) 1 μl dNTPs (10 mM each).
   (c) 0.5 μl Phusion hot start polymerase.
   (d) 10 μl 5× Phusion HF buffer.
   (e) 0.75 μl PCR primer 1.1 Illumina.

(f) 0.75 μl PCR primer 1.2 Illumina.

(g) 0.75 μl DMSO 1.5 PCR primer 1.1. Illumina.

Add H_2O to a final volume of 50 μl.

2. Amplify using the following PCR protocol:

   (a) 98 °C for 30 s.

   (b) 20 cycles of:

   - 98 °C for 10 s.
   - 65 °C for 30 s.
   - 72 °C for 30 s.

   (c) 72 °C for 5 min.

   (d) Hold at 4 °C.

3. Run 5 μl of PCR product on a 1.5 % agarose gel stained with Gel Red.
4. Purify DNA using the QIAquick PCR Purification Kit and elute with 12 μl of Elution Buffer.
5. Quantify by Qubit fluorometer (Invitrogen) using the sensitive assay Quant-iT dsDNA High Sensitive Assay kit. This method is not only quantitative; it is an advisable quality control analysis on the sample library using an Agilent Technologies 2100 Bioanalyzer.

4 Notes

1. Avoid the exposition to formaldehyde hazardous vapors using a chemical fume hood. Use high quality formaldehyde that is not past the manufacturer's expiration date. It is possible to prepare fresh 18.5 % formaldehyde from powdered paraformaldehyde to use immediately:

   (a) 4.8 ml of distilled water to a 50 ml tube.

   (b) 0.925 g paraformaldehyde.

   (c) 35 μl of 1 N KOH.

   (d) Place the tube in a 500 ml glass beaker filled with approximately 200 ml of water.

   (e) Microwave until water in beaker begins boiling.

   (f) Vortex the tube until paraformaldehyde begins to dissolve.

   (g) Repeat **steps e** and **f** until paraformaldehyde is completely dissolved.

   (h) Store on ice until cool and use immediately.

2. Agitation of cells is recommended. This step is a time-critical procedure; excessive cross-linking can mask the antibody recognition epitopes and result in a decrease in the amount of

bound to DNA-protein. The time of incubation can be increased to 30 min depending on the protein of interest, performing a time-course experiment is recommended to optimize the cross-linking conditions.

3. Glycine stops the cross-linking reaction by quenching the formaldehyde. A color change of the culture was observed due to the change in the pH of the medium by adding glycine.
4. At this point, the completely dry cell pellet can be frozen at –80 °C.
5. Sonication produces heat, which can denature the chromatin. Keep the cell lysate on ice cold and alternates cycles of sonication with breaks to avoid sample overheating. We recommend using low binding DNA tubes. Avoid sample foam during sonication.
6. For example, using a Bioruptor (Diagenode) instrument for HEL cell line, 500 μl of cell lysate in low binding 1.5 ml tubes with a cell density of $4–5 \times 10^7$/ml and SDS Sonication Buffer. Set the Bioruptor on high power setting. Subject the cells to two run of ten cycles: 30 s "ON," 30 s "OFF" each. Spin and mix with vortex between each run.
7. To save time in checking the sonication step, we can follow these two options:

   Option 1:

   - Add 10 μl of Elution Buffer to 10 μl of sample.
   - Incubate at 95 °C for 5 min.
   - Load 10 μl in a 1 % agarose gel, stained with Red Safe, and a 100 bp DNA marker.
   - Observe which of the shearing conditions gives a smear of DNA in the range of 200–1000 bp, *see* Fig. 2 for an example.

   Option 2:

   - Add 1 μl RNAse A (10 mg/ml) and incubate at 37 °C for 30 min.
   - Add 1 μl Proteinase K and incubate at 62 °C for 2 h.
   - Load 10 μl in a 1 % agarose gel, stained with Red Safe, and a 100 bp DNA marker.
8. If the sonication buffer does not contain SDS (Buffer 4), we should add 10× RIPA Buffer to obtain a 1× SDS final concentration because it is necessary for immunoprecipitation (e.g., prepare 1.5 ml low binding microtubes each containing 900 μl of sheared chromatin, for the number of desired immunoprecipitations, put on ice and add 100 μl of 10× RIPA Buffer containing the Protease Inhibitor Cocktail).

9. If multiple immunoprecipitations are performed from the same chromatin preparation, use the appropriate volume of Dilution Buffer containing Protease Inhibitor Cocktail for the correct number of immunoprecipitations in the same tube, a low binding tube when possible.
10. The preclearing step removes proteins or DNA from the cell lysate to prevent nonspecific binding of these components to the IP beads or antibody. The protein agarose beads should be pre-blocked with BSA.
11. Resuspend completely the protein beads slurry before adding to samples. Cut the end of the tip to pick up the agarose beads, mix by inverting, do not use vortex to avoid deforming them. Magnetic protein beads can also be used.
12. If multiple immunoprecipitation share performed from the same chromatin preparation, use the appropriate volume of agarose beads in the same tube.
13. If different chromatin preparations are carried together through this protocol, remove 1 % of the chromatin as Input for each.
14. The appropriate amount of antibody needs to be empirically determined, the typical amount is 5 μg, but can vary between 1 and 10 μg.
15. Add the same amount for both specific and Normal IgG antibodies.
16. When possible, ChIP-grade antibodies should be used. The effects of cross-linking can generate new epitopes and specific epitopes may be lost. It is preferable polyclonal antibodies where a number of antibodies recognize different epitopes. Polyclonal antibodies reduce the probability that all specific epitopes be masked by the cross-linking process.
17. The amount of DNA recovery depends on the antibody used. Usually range from 5 to 50 ng from 2×10^7 cells. Total chromatin for the control sample recovers about 20 μg.
18. The amount of adapter depends on the amount of DNA and should be tritiated.
19. The ligation time may be shorter.

Acknowledgments

This work was supported by the project "Efecto del Ácido Retinóico en la enfermedad alérgica. Estudio transcripcional y su traslación a la clínica," PI13/00564, integrated into the "Plan Estatal de I+D+I 2013–2016" and cofunded by the "ISCIII-

Subdirección General de Evaluación y Fomento de la investigación" and the European Regional Development Fund (FEDER).

We are grateful to members of our laboratory for their stimulating discussion of this review.

References

1. Aparicio JG, Viggiani CG, Gibson DG et al (2004) The Rpd3-Sin3 histone deacetylase regulates replication timing and enables intra-S origin control in Saccharomyces cerevisiae. Mol Cell Biol 24:4769–4780
2. Kouzarides T (2007) Chromatin modifications and their function. Cell 128:693–705
3. Kuo MH, Allis CD (1999) In vivo cross-linking and immunoprecipitation for studying dynamic Protein:DNA associations in a chromatin environment. Methods 19:425–433
4. Gilmour DS, Lis JT (1986) RNA polymerase II interacts with the promoter region of the noninduced hsp70 gene in Drosophila melanogaster cells. Mol Cell Biol 6:3984–3989
5. Gilmour DS, Lis JT (1985) In vivo interactions of RNA polymerase II with genes of Drosophila melanogaster. Mol Cell Biol 5:2009–2018
6. Gilmour DS, Lis JT (1984) Detecting protein-DNA interactions in vivo: distribution of RNA polymerase on specific bacterial genes. Proc Natl Acad Sci U S A 81:4275–4279
7. Solomon MJ, Larsen PL, Varshavsky A (1988) Mapping protein-DNA interactions in vivo with formaldehyde: evidence that histone H4 is retained on a highly transcribed gene. Cell 53:937–947
8. Hecht A, Strahl-Bolsinger S, Grunstein M (1996) Spreading of transcriptional repressor SIR3 from telomeric heterochromatin. Nature 383:92–96
9. Boyd KE, Farnham PJ (1997) Myc versus USF: discrimination at the cad gene is determined by core promoter elements. Mol Cell Biol 17:2529–2537
10. Parekh BS, Maniatis T (1999) Virus infection leads to localized hyperacetylation of histones H3 and H4 at the IFN-beta promoter. Mol Cell 3:125–129
11. Wathelet MG, Lin CH, Parekh BS et al (1998) Virus infection induces the assembly of coordinately activated transcription factors on the IFN-beta enhancer in vivo. Mol Cell 1:507–518
12. Hebbes TR, Thorne AW, Crane-Robinson C (1988) A direct link between core histone acetylation and transcriptionally active chromatin. EMBO J 7:1395–1402
13. Collas P (2010) The current state of chromatin immunoprecipitation. Mol Biotechnol 45: 87–100
14. Zeng PY, Vakoc CR, Chen ZC et al (2006) In vivo dual cross-linking for identification of indirect DNA-associated proteins by chromatin immunoprecipitation. Biotechniques 41: 694–698
15. O'Neill LP, Turner BM (2003) Immunoprecipitation of native chromatin: NChIP. Methods 31:76–82
16. O'Neill LP, VerMilyea MD, Turner BM (2006) Epigenetic characterization of the early embryo with a chromatin immunoprecipitation protocol applicable to small cell populations. Nat Genet 38:835–841
17. Hanlon SE, Lieb JD (2004) Progress and challenges in profiling the dynamics of chromatin and transcription factor binding with DNA microarrays. Curr Opin Genet Dev 14: 697–705
18. Sikder D, Kodadek T (2005) Genomic studies of transcription factor-DNA interactions. Curr Opin Chem Biol 9:38–45
19. Geisberg JV, Struhl K (2004) Quantitative sequential chromatin immunoprecipitation, a method for analyzing co-occupancy of proteins at genomic regions in vivo. Nucleic Acids Res 32:e151
20. Loh YH, Wu Q, Chew JL et al (2006) The Oct4 and Nanog transcription network regulates pluripotency in mouse embryonic stem cells. Nat Genet 38:431–440
21. Wie C, Wu Q, Vega VB et al (2006) A global map of p53 transcription-factor binding sites in the human genome. Cell 124:207–219
22. Dahl JA, Collas P (2007) A quick and quantitative chromatin immunoprecipitation assay for small cell samples. Front Biosci 12:4925–4931
23. Dahl JA, Collas P (2007) Q2ChIP, a quick and quantitative chromatin immunoprecipitation assay, unravels epigenetic dynamics of developmentally regulated genes in human carcinoma cells. Stem Cells 25:1037–1046
24. Flanagin S, Nelson JD, Castner DG et al (2008) Microplate-based chromatin immunoprecipitation method, Matrix ChIP: a platform

to study signaling of complex genomic events. Nucleic Acids Res 36:e17
25. Fullwood MJ, Han Y, Wei CL et al (2010) Chromatin interaction analysis using paired-end tag sequencing. Curr Protoc Mol Biol Chapter 21, p Unit 21 15 1-25
26. Johnson KD, Bresnick EH (2002) Dissecting long-range transcriptional mechanisms by chromatin immunoprecipitation. Methods 26: 27–36
27. Li G, Fullwood MJ, Xu H et al (2010) ChIA-PET tool for comprehensive chromatin interaction analysis with paired-end tag sequencing. Genome Biol 11:R22
28. Nelson JD, Denisenko O, Bomsztyk K (2006) Protocol for the fast chromatin immunoprecipitation (ChIP) method. Nat Protoc 1: 179–185
29. Nelson JD, Denisenko O, Sova P et al (2006) Fast chromatin immunoprecipitation assay. Nucleic Acids Res 34:e2
30. Orlando V (2000) Mapping chromosomal proteins in vivo by formaldehyde-crosslinked-chromatin immunoprecipitation. Trends Biochem Sci 25:99–104
31. Ren B, Dynlacht BD (2004) Use of chromatin immunoprecipitation assays in genome-wide location analysis of mammalian transcription factors. Methods Enzymol 376:304–315
32. Lee TI, Johnstone SE, Young RA (2006) Chromatin immunoprecipitation and microarray-based analysis of protein location. Nat Protoc 1:729–748
33. Barski A, Cuddapah S, Cui K et al (2007) High-resolution profiling of histone methylations in the human genome. Cell 129: 823–837
34. Johnson DS, Mortazavi A, Myers RM et al (2007) Genome-wide mapping of in vivo protein-DNA interactions. Science 316:1497–1502
35. Mikkelsen TS, Ku M, Jaffe DB et al (2007) Genome-wide maps of chromatin state in pluripotent and lineage-committed cells. Nature 448:553–560
36. Robertson G, Hirst M, Bainbridge M et al (2007) Genome-wide profiles of STAT1 DNA association using chromatin immunoprecipitation and massively parallel sequencing. Nat Methods 4:651–657

Chapter 10

Gene Silencing Delivery Methods: Lipid-Mediated and Electroporation Transfection Protocols

Asunción García-Sánchez and Fernando Marqués-García

Abstract

The RNA interference (RNAi) plays an important role in regulation of gene expression. It is a mechanism used by many organisms to silence the expression of genes that control different processes in the cell. The double strand (ds) RNA molecule inhibits gene expression of a targeted gene with high specificity and selectivity.

Different types of small ribonucleic acid molecules, microRNA (miRNA), small interfering RNA (siRNA), short hairpin RNA (shRNA), and the piwi RNA (piRNA) are involved in the RNA interference. RNAi is a relevant research tool in cell cultures and in vivo experiments because synthetic dsRNA introduced into cells can selectively silence specific target genes.

Here, we describe a general guide for gene silencing mediated by siRNA, focusing on the most used delivery methods: lipid-mediated and electroporation transfection.

Key words Electroporation transfection, Lipid-mediated transfection, miRNA, mRNA regulation RNA interference, shRNA, siRNA

1 Introduction

RNA interference (RNAi) is an essential mechanism for regulation of genetic information during plant and animal development [1]. Noncoding RNA molecule inhibits gene expression by destruction of specific mRNA molecules or blocking the transcription. RNAi was discovered in 1998, when Andrew Fire and Craig C. Mello described the ability of exogenous double-stranded RNAs (dsRNA) to silence genes specifically in *Caenorhabditis elegans* [2]. Years later, in 2006 they were awarded with Nobel Prize in Physiology or Medicine for their work on RNA interference.

RNA interference is activated when RNA molecule is present as double-stranded pairs in the cell. Double-stranded RNA (dsRNA) activates the RNAi pathway. It is initiated by the enzyme Dicer [1], which cleaves long dsRNA molecules into short double-stranded fragments of ~20 nucleotide RNAs and 3′-ends overhangs named siRNAs (short interfering RNA) [3]. The siRNA then

María Isidoro-García (ed.), *Molecular Genetics of Asthma*, Methods in Molecular Biology, vol. 1434,
DOI 10.1007/978-1-4939-3652-6_10, © Springer Science+Business Media New York 2016

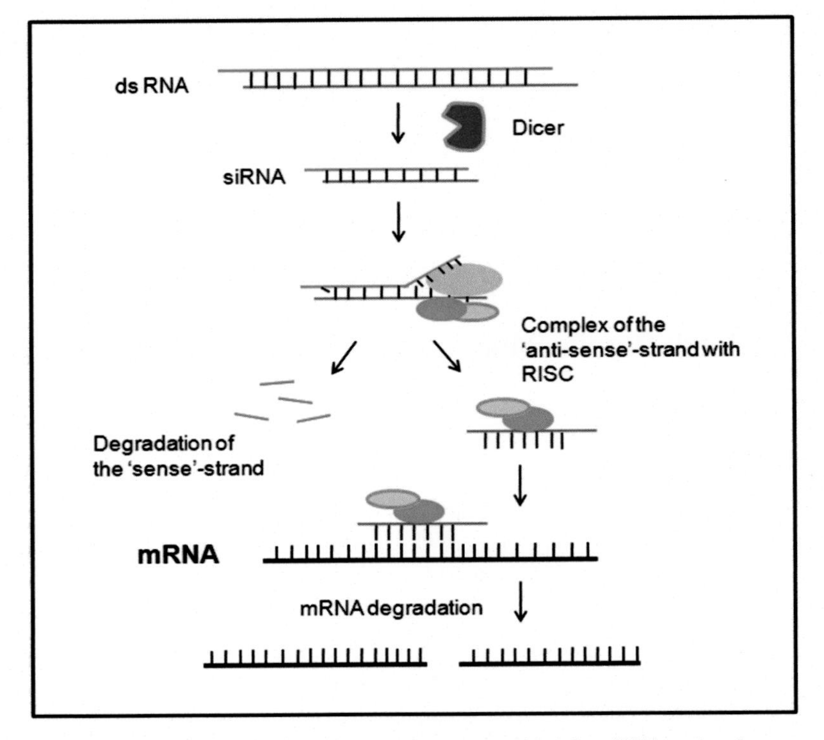

Fig. 1 RNAi mechanism and schematic representation of a siRNA molecule

associates with RISC (RNA-induced silencing complex), a large protein complex (160 kDa) with a catalytic component, the Argonaute proteins [4]. Inside the RISC, the siRNA is unwound into two single-stranded RNAs (ssRNAs), the passenger (sense) strand, and the guide (antisense) strand. The passenger strand is degraded by cellular nucleases and the guide strand directs RISC to the complementary target mRNA sequence. This mRNA target is degraded by Argonaute [5] (Fig. 1).

piRNA is the latest class of noncoding RNA identified, and its mechanism of biogenesis and function are not completely understood. One well-characterized piRNA function is the silencing of mobile elements [6].

RNAi is a relevant mechanism for the regulation of gene expression avoiding viral infections by knocking down viral mRNAs [7–9] and keeping jumping genes under control [10]. It is widely used in both basic and applied research as a method to study gene function and to develop promising gene therapies [11, 12].

Experimental RNAi in the mammalian system involves the introduction of siRNA, typically 21–23 bp duplexes, or shRNA that consists of a single strand having the sequence of the two desired siRNA strands connected by a no relevant sequence. The two complementary portions anneal intra-molecularly allowing the RNA

Table 1
Differences between siRNA and shRNA

	siRNA	shRNA
Nomenclature	Small interference RNA	Short hairpin RNA
Source	Laboratory synthesis	Nuclear expression
Delivery	Via lipid-mediated transfection, electroporation to the cytoplasm	Via viral-mediated transfection or DNA vector to the nucleus
Expression	Transient (48–72 h)	Stable (up to 3 years)

molecule to fold back on itself, creating a dsRNA molecule with a hairpin loop.

The Dicer enzyme cleaves the RNA structure into the desired siRNA molecule. The siRNA can be prepared for several methods, such as chemical synthesis (the same chemistry used to generate DNA primers for PCR), in vitro transcription, siRNA expression vectors (containing siRNA or shRNA), and PCR expression cassettes. Both siRNA and shRNA are intrinsically different molecules and present differences in the molecular mechanism of action and applications [13] (Table 1).

In this chapter, a protocol of designing and delivering siRNA in both adherent and suspension cell lines will be described.

2 Materials

2.1 Lipid-Mediated Transfection for Adherent Cells

1. Human HEK 293T embryonic cells or another cell line of interest.
2. DMEM Low Glucose, L-Glutamine, supplemented with 10% fetal bovine serum (FBS) and penicillin/streptomycin or another appropriate medium depending on the used cell line.
3. 1× PBS (Phosphate-Buffered Saline).
4. 1× Trypsin/EDTA: 0.05% porcine trypsin and 0.02% EDTA in Hank's balanced salt solution with phenol red.
5. 75 μM siRNA of interest.
6. 75 μM Silencer Select Negative Control #1 siRNA (Life Technologies, Thermo Fisher).
7. 75 μM Silencer Select GAPDH Positive siRNA Human (Life Technologies, Thermo Fisher).
8. 75 μM Silencer FAM-labeled Negative Control #1 siRNA (Ambion).
9. Lipofectamine 2000 transfection reagent (Invitrogen).

10. Opti-MEM I Reduced Serum Medium (Gibco).
11. Humidified 37 °C/5 % CO_2 incubator.
12. Laminar flow cell culture hood.
13. Six-well tissue culture plates.
14. Micropipettes.
15. Filter tips.

2.2 Electroporation

1. Cell line of interest.
2. Supplemented Nucleofector Solution at room temperature (Lonza).
3. pmaxGFP Vector (2 μg/nucleofection) (Lonza).
4. 50 pmol/nucleofection siRNA of interest.
5. Cell Line Optimization Nucleofector Kit (Lonza).
6. Culture medium appropriate for the cell of interest.
7. RPMI-1640.
8. Sterile 1× PBS (Phosphate Buffer Saline).
9. Laminar flow cell culture hood.
10. Humidified 37 °C/5 % CO_2 incubator.
11. Nucleofector 2b Device (Lonza).
12. 12-Well tissue culture plates.
13. Nucleofector cuvettes (Lonza).
14. Nucleofector plastic pipettes (Lonza).
15. Micropipettes.
16. Filter tips.

3 Methods

3.1 siRNA Design

Finding the functional binding site on a target mRNA sequence, which will correspond to the sense strand of the siRNA, is the first step in the designing siRNA. Currently, there are several algorithms available for designing siRNAs, and commercial companies specialized in that offer predesign siRNA with a guaranteed functionality. These commercial siRNAs result in a quick (a few days), highly effective, and straightforward approach. The disadvantage is its high cost.

On the other hand, the researchers can design their own siRNAs:

1. Find 21 nucleotide sequences in the target mRNA located 50–100 nt downstream of the start codon (AUG).
2. Search for sequence motif AA (*see* **Note 1**).

3. Select 2–4 target sequences.
4. Target sequences should have G + C content between 30 and 50 %.
5. Avoid stretches of four or more nucleotide repeats, especially T or A, because a 4–6 nucleotide poly (T) acts as a termination signal for RNA polymerase III.
6. Select the siRNA target sites at different positions along the length of the gene sequence (*see* **Note 2**).
7. Avoid sequences that share certain degree of homology with other related or unrelated genes. Compare the potential target sites to the genome database by BLAST (www.ncbi.nlm.nih.gov/BLAST). Do not consider any target sequences with more than 16–17 contiguous base pairs of homology to other coding sequences.
8. Design a negative control siRNA with the same nucleotide composition as your siRNA but lacking significant sequence homology to the genome. Mix the nucleotide sequence of your siRNA target and make a BLAST to check the lack of homology.

3.2 Use Appropriate Controls

Each experiment should include the following controls in triplicate:

1. Untreated cells are necessary to check the cytotoxicity of the siRNA delivery and normalize the target mRNA and protein levels both in control and in experimental samples.
2. Positive control with siRNA (endogenous siRNA or reporter gene) to optimize and monitor efficiency of siRNA delivers into cells.
3. Negative control siRNA (nontargeting) constitute a basal reference for target-specific knockdown gene.

A rescue experiment control consists in transfect cells with recombinant protein to re-introduce the protein [14]. Often is required by journal reviewers for publication.

3.3 Delivery Methods

A successful delivery of siRNA into cells is the most critical step for efficient and effective silencing of the expression of a target gene in cultured mammalian cells by RNAi. A poor delivery of siRNA is the most common reason for ineffective gene silencing. The delivery method of choice depends on the cell line. Not all reagents work well in all cell types and for all applications.

The optimization of any delivery method is necessary for every cell type and is advisable to try 2–4 different cell densities. In order to enter the cell, siRNA must come into contact with cell membrane. Three major technologies have been developing to introduce siRNA into target cells: lipid-mediated transfection, electroporation, and viral gene transfer (Table 2).

3.3.1 Lipid-Mediated Transfection of siRNA in Adherent Cell Line

Transfection usually involves the use of packaging particles named liposomes to facilitate the cellular entry of siRNA. The liposomes interact with nucleic acids and fuse with the cellular membrane allowing the entrance in the cell [15]. The lipid-mediated transfection is the method of choice for the majority of cell types, particularly any stable and secondary, transformed adherent cell line. It is necessary to optimize transfection conditions before beginning the siRNA experiment.

1. The day before the transfection, wash the cell monolayer with 1× PBS (*see* **Notes 3** and **4**).
2. Add 1× trypsin/EDTA (1 ml/25 cm² of surface area) to the cell monolayer. Allow the solution to cover the entire monolayer surface. Incubate the flask into the incubator at 37 °C 5 % CO_2 for 5–10 min (*see* **Note 5**).
3. Visualize the cells under an inverted microscope to check the detached state of the cells. Most of the cells should be floating (*see* **Note 6**).
4. Resuspend the cells in a small volume of fresh serum-containing medium to inactivate trypsin.
5. Remove the cell suspension to a new 15 ml tube and centrifuge at ≥370 × *g* for 5 min at room temperature to eliminate all trypsin.
6. Resuspend the cells in a small volume of antibiotic-free normal growth medium supplemented with FBS, remove a small volume and mix with the same volume of Trypan Blue for counting the viable cells in a Neubauer counter chamber.

Table 2
Different siRNA delivery methods

Technique	Delivery mode	Advantages	Disadvantages
Transfection	Cationic liposomes	Delivery of siRNA, miRNAs, and shRNAs into most cell types (typically adherent cell lines) High efficiency and reproducibility Many commercial formulations	Toxicity of reagents Not working in all cell types (primary and nondividing cells)
Electroporation	Electrical pulse	Effective for difficult-to-transfect cells (typically suspension cell lines) Delivery into difficult-to-transfect cells	High cell death Optimization for different cell types required
Viral-mediated delivery	Lentivirus Retrovirus Adenovirus	Effective for difficult-to-transfect cells. Stable selection In vivo application	Needed biosafety Level 2 Cabinet Possible mutagenic and immunogenic effects
Modified siRNA	Passive uptake	Effective for difficult-to-transfect cells In vivo application	Delivery efficiency inhibited by serum

7. In a six-well culture plate, seed 25×10^5 cells per well in 2 ml of antibiotic-free normal growth medium supplemented with FBS.
8. Incubate the cells at 37 °C in a 5% CO_2 incubator until the cells are 50–70% confluent (usually 18–24 h) (*see* **Notes 7** and **8**).
9. Prepare the siRNA oligomers at the appropriate concentration to use (*see* **Note 9**).
10. For transfection, prepare the following solutions:

    Solution A: For each transfection, dilute 2 μg of siRNA and controls into 250 μl Opti-MEM (**Note 10**).

    Solution B: For each transfection, dilute 5 μl Lipofectamine 2000 into 250 μl Opti-MEM. Incubate 5 min at room temperature (*see* **Note 11**).
11. Add solution B (diluted Lipofectamine 2000 Reagent) to each solution A (diluted siRNA) (1:1 ratio) (*see* **Note 12**).
12. Incubate at room temperature for 20–30 min.
13. Remove the culture medium of the cells.
14. Add siRNA-lipid complexes to cells drop by drop and gently move the plate from one side slowly to the other side (*see* **Note 13**).
15. Incubate the cells in a humidified 37 °C/5% CO_2 incubator.
16. After 4–6 h, remove the complexes and add fresh medium supplemented with antibiotics.
17. Fluorescein conjugated Control siRNA (Silencer FAM-labeled Negative Control #1 siRNA) should be incubated only for 5–7 h in a humidified 37 °C/5% CO_2 incubator. After this time, they are ready to be analyzed by flow cytometer or fluorescent microscopy.
18. Incubate cells for 24–48 h in a humidified 37 °C/5% CO_2 incubator until they are ready to analyze for gene knockdown.
19. After incubation, remove the medium, wash the monolayer, and trypsinize it as explained in **steps 2–5**.
20. Wash the cellular pellets with 500 μl of 1× PBS and centrifuge at $\geq 370 \times g$ for 5 min at room temperature.
21. The pellet of transfected cells with control siRNA (FAM-labeled) is resuspended in normal medium and ready to flow cytometer analyzing.
22. The pellets of transfected cells with siRNA and siRNA control are ready to RNA/protein extraction in order to analyze the inhibition of mRNA target gene (*see* **Note 14**).
23. To isolate RNA for subsequent RT-PCR analysis it is usefull the use of the RNeasy Mini Kit (Qiagen), which includes an on-column DNAse digestion with RNAse-free DNAse, that facilitates the elimination of possible genomic DNA contamination.

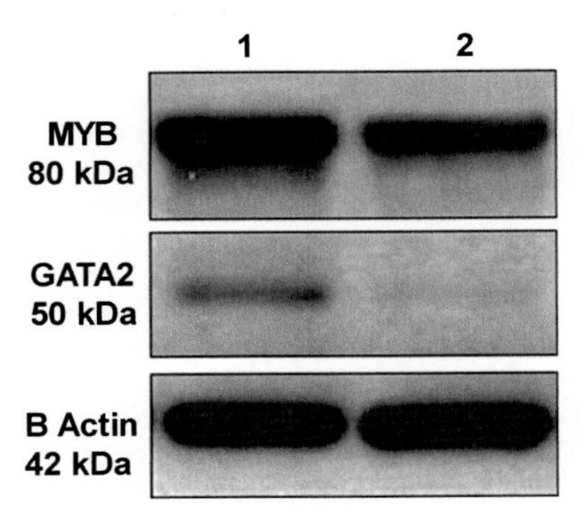

Fig. 2 Western Blot evaluation of siRNA knockdown. Hek293T cells were transfected with nontargeting or targeting siGATA2. Myb and β Actin serve as negative and loading control, respectively for the nontargeting and siGATA2

24. For Western blot analysis, prepare the nuclear lysates using Pre NE-PER Nuclear and Cytoplasmic Extraction Reagent or other method of convenience and proceed with Western blot after protein quantification (*see* **Note 15**) (Fig. 2).

3.3.2 Electroporation Transfection of siRNA in a Suspension Cell Line

Many cell types (such as primary cell lines or suspension cell lines among others) are difficult to transfect with any nucleic acid using lipids. For these cell lines, the method of choice is the electroporation consisting in a brief but powerful electric pulse during which lipid molecules of cell membrane are disrupted allowing the entry of nucleic acid. However, this method causes a high rate of cell death [16].

1. Visit the web page http://bio.lonza.com/ to check if the target cell line is optimized for the technical equipment (*see* **Note 3**).
2. If the target cell line is optimized, use the recommendations for the company. If not, use the "Cell Line Optimization Nucleofector Kit" to determine the optimal conditions to transform the cell line of interest (*see* **Note 16**).
3. One Nucleofector sample contains 2 μg pmax GFP as positive control, 100 μl Nucleofector Solution, and 5×10^6 of cell suspension.
4. Add 18 μl of nucleofector supplement to 82 μl of nucleofector solution (both stored at 4 °C) for each nucleotransfection. Mix by pipetting up and down. Pre-warm the mixture to room temperature, scale up the amount of nucleofector supplement solution according to the total (*see* **Note 17**).
5. Prepare four 12-well culture plates and fill 9 of them with 1 ml of fresh supplemented culture medium and incubate them in a 37 °C/5 % CO_2 incubator (*see* **Note 18**).

6. Count the cells and determine cell density.
7. Transfer the appropriate amount of cells for the nine nucleofector assays (4.5×10^7) in two 1.5 ml Eppendorf tubes with solution L and other two for the assays with solution V.
8. Centrifuge the samples at $100 \times g$ for 10 min at room temperature. Remove supernatant completely.
9. Resuspend each cell pellet carefully in 900 μl of Nucleofector-supplemented solution (*see* **Note 19**).
10. Mix 800 μl of cell suspension (8 samples) with 16 μg pmaxGFP Vector. Keep 100 μl of cell suspension without DNA (*see* **Note 20**).
11. Transfer 100 μl of each aliquot into cuvettes and close it with the cap (*see* **Note 21**).
12. Select the Nucleofector Program, insert the cuvette with cell/DNA suspension into the Nucleofector, and press the button (*see* **Note 22**).
13. Take the cuvette out of the nucleofector when the program is finished.
14. Add ~500 μl of the warmed culture medium to the cuvette and gently transfer the sample drop by drop into the prepared 12-well plates (final volume ~1.5 ml media per well) (*see* **Notes 23–25**).
15. Incubate the cells in 37 °C/5% CO_2 incubator until analysis.
16. Check the transfection efficiency and cell viability by flow cytometer at 24 h after transfection.
17. Once the optimization assay is performed and the proper settings are known, proceed to the nucleofection experiment with the siRNA of interest, that facilitates the elimination of possible genomic DNA contamination.
18. Transfected cells with siRNA and siRNA control are ready to RNA/protein extraction in order to analyze the inhibition of mRNA target gene (*see* **Note 13**).
19. To isolate the RNA for subsequent RT-PCR analysis it is usefull the use of the RNeasy Mini Kit, which includes an on-column DNAse digestion with RNAse-free DNAse.
20. For Western blot analysis, prepare the cell nuclear lysates using Pre NE-PER Nuclear and Cytoplasmic Extraction Reagent or other method of convenience and proceed with Western blot after protein quantification (Fig. 2) (*see* **Note 15**).

The knockdown of target genes mediated by siRNA transfection is usually temporary, known as transient transfections. The synthetic shRNA is developed to achieve a stable or long-term inhibition of a target gene. The delivery of this RNAi approach is mediated by recombinant viral vectors, based on retrovirus, adenovirus, adeno-associated virus, and lentivirus DNA. These viral vectors permit a stable siRNA expression and the nucleic acid entry in cells where other methods such as lipofection do not work [17].

4 Notes

1. This strategy for choosing siRNA target sites is based on the Elbashir et al.'s observations [3].
2. Avoid 5′ UTR and 3′ UTR because these regions contain binding sequences for regulatory proteins that may affect the accessibility of the RNA target sequence to the RISC complex, although siRNAs targeting UTRs have successfully induce gene silencing.
3. When possible, use a cell line that is known to have high transfection efficiency. Usually, primary cells are not easily transfected. Cells should be 30–70 % confluent and at a relative low passage number (e.g. <50). Old cell cultures may present different phenotype and growth characteristics compared with those cells from early passages. Overgrown cultures or with too low densities may cause different expression profiles that will affect the transfection experiments. To ensure reproducibility, it is important to maintain a similar passage number between the experiments.
4. Use approximately half volume of culture medium of PBS. The serum has to be completely removed before adding trypsin.
5. Avoid prolonged incubations with trypsin because they can damage the cell surface receptors.
6. Gently hit the walls of flask to allow the cells to be released.
7. Cell density affects the transfection efficiency. Do not use 90 % confluent cells.
8. Transfecting cells at a lower density permits longer incubation times between transfection and analysis, avoiding the loss of cell viability due to cell overgrowth. The cell densities may be optimized, depending on the gene target.
9. Use high quality siRNA. The purification of siRNA should be correctly done avoiding reagents such as ethanol, salts, or other contaminants that result in cytotoxicity. Use the low effective concentration of siRNA recommended with the delivery method used in order to minimize nonspecific effects. It is advisable to perform a dose response curve to optimize the amount of siRNA or plasmid to not over activate the RISC complex or induce toxicity in the cell.
10. Use appropriate culture media; it is reported that antibiotics may increase toxicity during siRNA delivery. In addition, some

siRNA transfection reagents require serum-free medium. Following the manufacturer instructions, RNA-Lipofectamine complexes must be made in serum-free medium such as Opti-MEM reduced Serum Medium.

11. Scale-up all the reactions in the same tube.
12. As a control of transfection efficiency, prepare the Silencer FAM-labeled Negative Control #1 siRNA in duplicate, one with and the other without Lipofectamine and measure the internalization of RNA inside the cell by the flow cytometer and the microscope.
13. Be careful, do not allow the cells to dry out before adding siRNA-lipid complexes.
14. Quantification of knockdown efficiency is performed by RT-PCR to check for presence of the mRNA target. It is recommended to measure each siRNA effect on the mRNA target performing a RT-qPCR before examining the effect of siRNA on protein levels. The reason is that the siRNA can effectively inhibit mRNA levels of the target gene, but it may not have this effect on protein levels if the target protein has a long half-life. However, Western Blot indicates the presence or absence of specific proteins (Fig. 2). In addition, this method can use antibodies to detect the effect that the knockout of the target protein has on other proteins. The presence or absence of the knockdown protein is observed using several staining that check the effect on expression and cellular location of other proteins. In order to know about the biological effect of target gene, inhibition, viability, and proliferation analyses can be performed.
15. Off-targeted effects occur when a siRNA is processed by the RISC complex and downregulates nonspecific target genes [18]. It is reported that this siRNA-induced toxicity is sequence-dependent, concentration-dependent, and target-independent [19]. miRNA and siRNA regulate mRNA through RNA interference. Depending on the level of complementarity between the siRNA and mRNA, this mRNA will be silenced (completely, via siRNA pathway on-targeted silencing) or transiently repressed (partial, via miRNA-like pathway). siRNA off-target effects occur through partial complementarity of the siRNA with the mRNA targets. Chemical modifications in seed region of siRNAs enhance antisense specificity and reduce the off-target effect [18]. Another approach to minimize off-targeting problems is using siRNAs pool at low concentration, which target one mRNA at several sites [20]. In this way, the unspecific effect on other mRNAs should be negligible [21, 22].
16. The Cell Line Optimization Nucleofector Kit contains two different solutions, V and L, which should be tested in combination with seven different Nucleofector Programs.

17. Add the supplement to the nucleofector solution just before the use. The nucleofector solution and nucleofector supplement can be stored at 4 °C for 6 months. Once supplemented, the solution can be stored at 4 °C for up to 3 months. Do not use the mixture beyond this period, as the transfection efficiency and cell viability are greatly reduced after this time.
18. We have found that transfection efficiency works best when cells are electroporated in antibiotic-free media.
19. The Nucleofector Solution should be at room temperature. Avoid leaving the cells more than 15 min in the Nucleofector Solution because it may reduce cell viability and transfection efficiency.
20. Cells should be resuspended in nucleofector solution/supplement mixture before the plasmid is added to avoid cell aggregates, which can result in reduced cell viability.
21. The sample must cover the bottom of the cuvette.
22. Be careful with air bubbles, as they interfere with the flow of electrical currents during nucleofection and thus result in reduced transfection efficiencies.
23. Avoid repeat aspiration of the sample.
24. The white foam in the cuvette indicates high cell death. Avoid catching them.
25. Cell viability is increased when pre-warmed 500 μl of RPMI medium is added to the cuvette immediately after nucleofection and gently transferred to a 1.5 ml Eppendorf tube. Incubate it 5–10 min in the 37 °C/5 % CO_2 incubator and then transfer it to the pre-warmed culture plates.

Acknowledgments

This work was supported by grants of the Junta de Castilla y León ref. GRS1047/A/14, GRS1189/A/15, and BIO/SA73/15; and by the project "Efecto del Ácido Retinóico en la enfermedad alérgica. Estudio transcripcional y su traslación a la clínica", PI13/00564, integrated into the "Plan Estatal de I+D+I 2013–2016" and cofunded by the "ISCIII-Subdirección General de Evaluación y Fomento de la investigación" and the European Regional Development Fund (FEDER).

References

1. Bernstein E, Caudy AA, Hammond SM et al (2001) Role for a bidentate ribonuclease in the initiation step of RNA interference. Nature 409:363–366
2. Fire A, Xu S, Montgomery MK et al (1998) Potent and specific genetic interference by double-stranded RNA in Caenorhabditis elegans. Nature 391:806–811

3. Elbashir SM, Lendeckel W, Tuschl T (2001) RNA interference is mediated by 21- and 22-nucleotide RNAs. Genes Dev 15: 188–200
4. Hammond SM, Bernstein E, Beach D et al (2000) An RNA-directed nuclease mediates post-transcriptional gene silencing in Drosophila cells. Nature 404:293–296
5. Martinez J, Tuschl T (2004) RISC is a 5′ phosphomonoester-producing RNA endonuclease. Genes Dev 18:975–980
6. Weick EM, Miska EA (2014) piRNAs: from biogenesis to function. Development 141: 3458–3471
7. Lecellier CH, Dunoyer P, Arar K et al (2005) A cellular microRNA mediates antiviral defense in human cells. Science 308:557–560
8. Gitlin L, Karelsky S, Andino R (2002) Short interfering RNA confers intracellular antiviral immunity in human cells. Nature 418: 430–434
9. Gitlin L, Andino R (2003) Nucleic acid-based immune system: the antiviral potential of mammalian RNA silencing. J Virol 77:7159–7165
10. Sijen T, Plasterk RH (2003) Transposon silencing in the Caenorhabditis elegans germ line by natural RNAi. Nature 426:310–314
11. Aigner A (2006) Gene silencing through RNA interference (RNAi) in vivo: strategies based on the direct application of siRNAs. J Biotechnol 124:12–25
12. Aigner A (2007) Applications of RNA interference: current state and prospects for siRNA-based strategies in vivo. Appl Microbiol Biotechnol 76:9–21
13. Rao DD, Senzer N, Cleary MA et al (2009) Comparative assessment of siRNA and shRNA off target effects: what is slowing clinical development? Cancer Gene Ther 16:807–809
14. Editorial (2003) Whither RNAi? Nat Cell Biol 5:489–490
15. Felgner PL, Gadek TR, Holm M et al (1987) Lipofection: a highly efficient, lipid-mediated DNA-transfection procedure. Proc Natl Acad Sci U S A 84:7413–7417
16. Tsong TY (1991) Electroporation of cell membranes. Biophys J 60:297–306
17. Shi Y (2003) Mammalian RNAi for the masses. Trends Genet 19:9–12
18. Jackson AL, Linsley PS (2010) Recognizing and avoiding siRNA off-target effects for target identification and therapeutic application. Nat Rev Drug Discov 9:57–67
19. Fedorov Y, Anderson EM, Birmingham A et al (2006) Off-target effects by siRNA can induce toxic phenotype. RNA 12:1188–1196
20. Straka M, Boese Q (2010) Current topics in RNAi: why rational pooling of siRNA is SMART. Thermo Fisher Scientific Inc., Waltham, MA, USA
21. Arvey A, Larsson E, Sander C et al (2010) Target mRNA abundance dilutes microRNA and siRNA activity. Mol Syst Biol 6:363
22. Larsson E, Sander C, Marks D (2010) mRNA turnover rate limits siRNA and microRNA efficacy. Mol Syst Biol 6:433

Chapter 11

Protocols for Exosome Isolation and RNA Profiling

Fernando Marqués-García and María Isidoro-García

Abstract

Exosomes are small extracellular vesicles of multivesicular bodies derived from the cell endosome. Molecules have a characteristic composition both in the membrane and in the carried load. These particles are located in the culture medium of cells in vitro and in vivo body fluids, being synthesized by almost all kind of cells. The most characteristic molecule transported within is the RNA with a large clinical potential. In this chapter, the methodology for isolation of exosomes and subsequent purification of the RNA is reviewed.

Key words Biomarkers, Exosome, Extracellular vesicles, Isolation, Precipitation, RNA profiling, Ultracentrifugation

1 Introduction

Cells synthesize molecules with which they perform specific metabolic functions. These molecules are functionally grouped, forming subcellular organelles, whose composition is different and variable in function of the physiological state of the cell and the type of organelle. Most are intracellular, but some are released into the extracellular space [1]. The discovery of extracellular vesicles released by the cells was done by the group of Stahl and Johnstone, in 1983, in studies of the reticulocytes [2]. Different types of vesicles have been described since then: exosomes, microvesicles, and apoptotic bodies, which differ in their cellular origin, physicochemical characteristics, and composition (Table 1).

In this chapter, we will focus only on the study of one type of these extracellular vesicles, the exosomes. Exosomes are small particles (40–100 nm) generated by the majority of culture cell lines and cells forming tissues in vivo, such as T lymphocytes [3], B lymphocytes [4], dendritic cells [5], or in pathological conditions (e.g., tumor cells) [6].

Depending on their cellular origin, the populations of exosomes differ, as cardiosoma derived of cardiomyocytes [7] or prostasome from prostate tissue [8]. Even, they can be grouped based

María Isidoro-García (ed.), *Molecular Genetics of Asthma*, Methods in Molecular Biology, vol. 1434,
DOI 10.1007/978-1-4939-3652-6_11, © Springer Science+Business Media New York 2016

Table 1
Extracellular vesicles types. Physical and chemical characteristics of the different extracellular vesicles groups.

	Exosomes	Microvesicles	Apoptotic bodies
Size	30–100 nm	100–1000 nm	50–500 nm
Density	1.10–1.21 g/ml	ID	1.16–1.28 g/ml
Sedimentation	100,000–200,000 g	10000–20,000 g	1200 g, 10,000–100,000 g
Form	Cup-shaped	Various shapes	Heterogeneous
Lipidic composition	Lipidic rafts[a]	Cholesterol and fosfatidilserin	ID
Proteins markers	CD63, CD9, CD81, Alix, Tsg101	Selectins, integrins, CD40	Histones
Intracellular origin	MVBs	Plasma membrane	ID
Release mode	Constitutive and regulate	Regulate	Regulate
Release mechanism	Exocytosis of MVBs	Plasma membrane budding	Cell shrinkage and death
Charge molecules	Proteins, ncRNA, mRNAs	Proteins, ncRNA, mRNAs	Proteins, ncRNA, mRNAs, DNA

ID indeterminate, *MVNs* multivesicular bodies, *ncRNAs* no codifing RNAs, *mRNAs* messenger RNAs
[a]Cholesterol, Ceramide, and Sphingomyelin

on their function as tolerasomes or oncosomes [9]. Also, in multiple biological fluids, for example, plasma [10], urine [11], milk [12], saliva [12], or BALF [13] can be found in a large amount (10^8–10^{11} particles/ml) [14], produced by both normal and pathological tissues; hence represent an important source of potential biomarkers.

Its basic function is the intercellular communication, transferring proteins, RNAs, lipids and DNA, from one cell to another [15–18]. Thus, this molecular information can change the physiology of the host cell at transcriptional, posttranscriptional, and epigenetic level. Exosomes are involved in different biological processes, depending on cell origin [19], such as the regulation of immune response, antigen presentation [20], apoptosis, angiogenesis, inflammation, coagulation [21], morphogen transport [22], spread of neurodegenerative diseases [23], and in processes of tumor genesis [24].

In the context of allergic disease, the exosomes can behave in two opposing ways [25]:

- Compounding the allergic response by acting as transport vehicles for allergen.

- Inducing tolerance to allergens. This is carried out by a subpopulation of exosomes, generated mainly in the intestinal epithelium, denominated tolerasomes.

These particles have also been proposed for therapy, by acting as vehicles and their potential as vaccines against infectious agents, vaccines for allergic diseases, and for treating autoimmune diseases [26].

Exosomes carry different types of molecules, among others proteins and RNA. One of the most abundant RNAs are the noncoding RNAs and messenger RNAs [27], which mark the origin of the exosome and have important functions on the host cell. RNA exosomes appears as a molecular signature for the diagnosis and prognosis of disease, since differences in the exosome RNA profiling between patients and controls have been found [28]. Thus, the study of the RNA content in these vesicles becomes very interesting [19].

Different proteins and RNA molecules identified in the exosomes have generated an enormous amount of information, which has been compiled in free high-capacity databases and are available via web for systematic analysis, highlighting two, ExoCarta (http://www.exocarta.org) [29] and EVpedia (http://evpedia.info/) [30]. The purpose of this chapter is to provide an overview of the currently employed methods for isolating exosomes as well as purification of RNA molecules contained therein. This chapter is completed by briefly describing the methods for determining the quality of the obtained exosomes.

2 Materials

The description of the material components are grouped into the following sections:

- Exosome isolation.
- Determination of the exosome quality.
- RNA isolation.

The exosomes can be obtained from two fundamental sample types:

- In vitro cell culture.
- Biological Fluids.

The exosomes can be obtained from supernatants of cell cultures (*see* **Note 1**).

The biological fluids are extracted in different types of tubes with or without additives according to the type of liquid (*see* **Note 2**). When working with blood samples, it is advisable using serum instead of plasma. Plasma samples are worst because they contain a high concentration of coagulation factors, which may co-precipitate with the exosomes. Extraction of exosomes can be done the day of sample collection or samples can be frozen at −80 °C until use (*see* **Note 3**).

2.1 Exosome Isolation

2.1.1 Exosome Isolation by Immunoisolation

1. Magnetic beads, Anti-EpCAM antibodies coupled to magnetic microbeads (DynaBeads, Invitrogen).

2.1.2 Exosome Isolation by Size-Exclusion Chromatography

1. Size-exclusion chromatography. Size-exclusion chromatography gel comprising 2 % agarose (Agarose Bead Technologies).

2.1.3 Exosome Isolation by Ultracentrifugation

1. 1× PBS (Phosphate-Buffered Saline).
2. Refrigerated centrifuge.
3. 50-ml polypropylene centrifuge tubes.
4. Ultracentrifuge and fixed-angle or swinging-bucket rotor.
5. Polyallomer tubes or polycarbonate bottles, rotor appropriate for the ultracentrifuge.
6. 0.22-μm filter sterilization device (Millipore).

2.1.4 Exosome Isolation by Precipitation

1. Total Exosome Isolation (from serum) (Invitrogen).
2. Total Exosome Isolation (from culture media) (Invitrogen).
3. 1× PBS (Phosphate-Buffered Saline).

2.2 Determination of Exosome Quality

2.2.1 Electron Microscopy

1. Uranyl acetate (Sigma-Aldrich).
2. 2.5 % Glutaraldehyde (Sigma-Aldrich).
3. 2 % Paraformaldehyde (Sigma-Aldrich).
4. Tetraspanins Antibodies (Abcam).
5. Formvar-carbon coated EM grids (Polysciences Inc.).

2.2.2 Flow Cytometry

1. 1× PBS (Phosphate-Buffered Saline).
2. MES (Sigma-Aldrich).
3. 200 mM Glycine (Sigma-Aldrich).
4. Tetraspanins Antibodies (Abcam).

2.3 RNA Isolation

1. Total Exosome RNA and Protein Isolation Kit (Invitrogen).
2. 14.3 M 2-mercaptoethanol.
3. 100 % Ethanol.
4. 1× PBS (Phosphate-Buffered Saline).
5. Microcentrifuge capable of at least 10,000 × *g*.

6. Heat block set to 95–100 °C.
7. RNase-free 1.5 ml or 2.0 ml polypropylene microfuge tubes, adjustable pipettes, and RNase-free tips.
8. Agilent 2100 Bioanalyzer (Agilent).

3 Methods

3.1 Exosome Isolation

At present, four different methods to obtain exosomes are used:

- Immunoisolation: based on the use of specific antibodies attached to magnetic beads.
- Size-Exclusion Chromatography: using specific columns for the retention of exosomes.
- Ultracentrifugation: two-step procedure of centrifugation at high speed.
- Precipitation or One-Step exosome Isolation: one-step procedure using specific reagents.

Of these, the last two, ultracentrifugation and precipitation, are the most widely used and will be described in greater detail in this chapter (*see* **Note 4**).

3.1.1 Exosome Isolation by Inmunoisolation

With this technique, total exosomes or more selective populations can be isolated based on specific markers of cell line (*see* **Note 5**).

1. The specific antibodies stick to the surface of a magnetic ball and capture the exosomes present in the sample (*see* **Notes 6** and **7**). This mixture is incubated for 2 h with stirring at RT.
2. Place the tubes in the magnetic separator and remove the fluid, leaving the magnetic beads and bound exosomes attached to the side of the tube.
3. Remove the tube from the magnetic separator, rinse the beads with PBS, and repeat the separation.
4. After the wash step, remove the tube from the magnetic holder and use the bead/exosome complex for RNA extraction.

3.1.2 Exosome Isolation by Size-Exclusion Chromatography

For isolation by size-exclusion chromatography, apply 2 ml of biological fluid (e.g., patient-derived cell-free ascites) to a 2% agarose-based gel column (2.5 × 16 cm). For optimal separation, the sample volume should be 1/20 of the total column volume.

1. Isocratically, elute the column with 1× PBS at a flow rate of 1 ml/min, monitoring at 280 nm and collect fractions (2 ml).
2. Pool the void volume fractions (based on absorbance at 280 nm) and centrifuge at 100,000 × *g* for 1 h at 4 °C.
3. Use the resulting pellet for extraction and analyze the RNA and/or proteins.

3.1.3 Exosome Isolation by Ultracentrifugation

Two protocols based on the type of sample, one for isolating exosomes from cell culture media (Protocol 1) and one for biological liquids (Protocol 2), will be described.

Protocol 1

1. Collect the culture medium in a centrifuge tube.
2. Centrifuge the medium, 20 min at 2000 × *g*, 4 °C, then eliminate cells and debris from the sample.
3. Collect the supernatant and transfer it to a polycarbonate tube, which is suitable for use in ultracentrifuge rotors.
4. Mark the tube centrifugation, at the bottom, guiding the brand toward the outside of the rotor. In this way, we ensure the position to locate the pellet and not lose it when removing the supernatant.
5. Centrifuge 30 min at 10,000 × *g*, 4 °C (*see* **Note 8**).
6. Collect the supernatant, taking care not to contaminate it with the pellet.
7. Centrifuge the obtained supernatant, 70 min at 100,000 × *g*, 4 °C (*see* **Note 9**).
8. Collect the supernatant completely (*see* **Note 10**).
9. Resuspend the pellet obtained in a high volume of 1× PBS (2–3 ml) (*see* **Note 11**).
10. As an additional purification step, you can filter the sample by using 0.22 μm filters.
11. Centrifuge 1 h at 100,000 × *g*, 4 °C.
12. Wash the pellet with 1 ml of 1× PBS, two times by repeating centrifugation conditions of **step 11** (*see* **Note 12**).
13. Resuspend the resulting pellet in a small volume of 1× PBS (50–100 μl) or in the commercial reagent of the RNA extraction kit.
14. Perform an extra concentration of exosomes starting in the supernatant from **step 11**. Centrifuge the supernatant 1 h at 100,000 × *g*, 4 °C. Remove most of the PBS visible above the pellet and resuspend exosomes in 20–50 μl of 1× PBS.
15. The sample can be stored for more than 1 year at −80 °C (aliquots of 100 μl are recommended).

Protocol 2

1. Centrifuge the sample of biological fluid (e.g., serum) 30 min at 2000 × *g*, 4 °C.
2. Collect the supernatant and transfer it to a polycarbonate tube, which is suitable for use in ultracentrifuge rotors.
3. Mark the tube centrifugation, as in **step 4** of Protocol 1.

4. Collect the supernatant and centrifuge 45 min at 12,000 ×*g*, 4 °C (*see* **Note 8**).
5. Collect the supernatant, taking care not to contaminate it with the pellet.
6. Centrifuge the supernatant 2 h at 110,000 ×*g*, 4 °C (*see* **Note 9**).
7. Collect the supernatant completely (*see* **Note 10**).
8. Resuspend the obtained pellet in a high volume of 1× PBS (2–3 ml) (*see* **Note 11**).
9. As an additional purification step, you can filter the sample using 0.22 μm filters.
10. Centrifuge 70 min at 110,000 ×*g*, 4 °C.
11. Wash the pellet with 1 ml of 1× PBS two times by repeating centrifugation conditions of **step 10** (*see* **Note 12**).
12. Resuspend the resulting pellet in a small volume of 1× PBS (50–200 μl) or in the commercial reagent of the RNA extraction kit.
13. Extra concentration of exosomes starting in the supernatant from **step 11**. Centrifuge the supernatant 1 h at 100,000 ×*g*, 4 °C. Remove most of the PBS visible above the pellet and resuspend exosomes in 20–50 ml PBS.
14. The sample can be stored for more than 1 year at −80 °C (aliquots of 100 μl are recommended).

3.1.4 Exosome Isolation by Precipitation

1. After thawing the sample, centrifuge 2000 ×*g* 30 min. In this way, we eliminate the debris, getting a free cell suspension (*see* **Note 13**).
2. Mix the free cell suspension with precipitating reagent (Total Exosome Isolation) in a ratio of 1 part to 5 parts reactive/sample (*see* **Note 14**).
3. Mix well by vortexing or pipetting up and down, until obtaining a smooth and clear mixture.
4. Incubate the sample 30 min 4 °C without agitation (*see* **Note 15**).
5. Centrifuge the sample 10 min 10,000 ×*g* RT (*see* **Note 16**).
6. Discard the supernatant and resuspend pellet in 1× PBS (100 μl) (*see* **Note 17**).
7. Keep the sample for 1 week at 4 °C, or store at −20 °C for a longer time (*see* **Notes 17–20**).

3.2 Determination of Exosome Quality

Before isolating the RNA molecules, it is advisable to determine the quality of the isolated exosome. Different techniques can be used:

3.2.1 Electron Microscopy

1. Place a drop, approximately 10 μg, of exosomal protein of the intact exosomes resuspended in PBS (5 μl) on Formvar-carbon coated EM grids. Prepare two or three grids for each exosome preparation. Cover and let the membranes adsorb for 20 min in a dry environment.
2. Put 100μl drops of PBS on a sheet of Parafilm. Transfer the grids (membrane side down) to drops of PBS with clean forceps to wash.
3. Transfer the grids to a 50μl drop of 1% glutaraldehyde for 5 min.
4. Transfer a 100μl drop of distilled water and let grids stand for 2 min. Repeat seven times for a total of eight water washes.
5. Immunostain with an appropriate antibody. Commonly used antibodies are anti-CD63 and anti-MHC class II.
6. Transfer the grid to a 30 μl drop of the primary antibody of choice and incubate for 40 min. Wash by repeating the **step 4**, but use 0.1% bovine serum albumin in PBS instead of PBS alone.
7. Repeat **step 5**, but with the 10 nm-gold labeled secondary antibody and wash with PBS alone.
8. Post-fix the sample by adding a drop of 2.5% glutaraldehyde to the Parafilm and incubate the grid on top of the drop for 10 min. Repeat the wash in point 4, but use five droplets of deionized water instead of three droplets of PBS.
9. Contrast the sample by adding a drop of 2% uranyl acetate to the Parafilm and incubate the grid on top of the drop for 15 min.
10. Embed the sample by adding a drop of 0.13% methylcellulose and 0.4% uranyl acetate to the Parafilm and incubate the grid on top of the drop for 10 min.
11. Remove excess liquid gently by using an absorbing paper, before positioning the grid on a paper with the coated side up and let it air dry for 5 min.
12. Examine the preparations with an electron microscope or store the grids in a grid box for future work.

3.2.2 Flow Cytometry

Depending on the exosomal cellular origin, different antibodies can be coupled to beads, magnetic or latex character (e.g., anti-MHC class II or anti-CD63-coated beads).

1. Wash 25 μl of 4 μm latex beads (30 × 106 beads) twice in 100 μl MES buffer, centrifuge at 3000 × *g* for 15–20 min, and dissolve the pellet in 100 μl MES buffer. Prepare the antibody mixture containing a volume equal of 12.5 μg antibody with the same volume of MES buffer. Add the beads to the

antibody mixture and incubate under agitation over night at room temperature. Wash the antibody-coated beads three times with 1× PBS (3000 × *g* for 20 min) and dissolve the pellet in 100 μl of storage buffer (with a final concentration of 300,000 beads/μl).

2. For each sample (each antibody), continue with a volume equal to a minimum of 30 μg of exosomal protein (of the intact exosomes solved in PBS) per ~100,000 antibody-coated beads.
3. Incubate exosomes and beads, in a total volume of 300 μl 1× PBS, overnight at 4 °C under gentle movement.
4. Block by adding 300 μl of 200 mM glycine and incubate for 30 min.
5. Wash the exosome-bead complexes twice in wash buffer (1–3 % serum in PBS), 600 × *g* for 10 min.
6. Incubate the exosome-bead complexes with 50 μl IgG antibody at 4 °C. Wash the exosome-bead complexes twice in wash buffer as described in **step 5**.
7. Add 90 μl of wash buffer and 10 μl of the antibody of choice (ideally anti-CD9, anti-CD63, or anti-CD81) to the exosome-bead complexes and incubate for 40 min, under gentle movement. Wash the exosome-bead complexes two times in wash buffer as described in **step 5**.
8. Add 300 μl wash buffer and acquire data using flow cytometry.

Western Blotting can also be employed to determine the quality of the isolated exosome. Dissolve the exosome pellet in the lysis buffer of choice and pipette thoroughly, followed by vortex mixing. To further lyse the exosomes, sonicate the sample in a water bath 3 × 5 min with vortex mixing in between. Finally centrifuge the sample, 13,000 × *g* for 5 min at room temperature, transfer the supernatant to a new eppendorf tube, and measure the total protein by a method of choice and load 20–50 μg of protein per well. The sample is then ready for western blotting.

3.3 RNA Isolation

1. From a volume of 200 μl of exosomes (if this volume is not available, complete with 1× PBS to that level), start to make RNA extraction (*see* **Notes 21** and **22**).
2. Add one volume of 2× Denaturing Solution and mix thoroughly (*see* **Notes 23** and **24**).
3. Incubate the mixture on ice for 5 min.
4. Add one volume of Acid-Phenol:Chloroform to each sample (*see* **Note 25**).
5. Mix samples by vortex for 30–60 s.

6. Centrifuge for 5 min at maximum speed (≥10,000 × *g*) at room temperature to separate the mixture into aqueous and organic phases. Repeat the centrifugation if the interphase is not compact.
7. Carefully remove the aqueous (upper) phase without disturbing the lower phase or the interphase, and transfer it to a fresh tube. Note the volume recovered. Final RNA isolation is performed on the aqueous phase from acid-phenol: chloroform extraction (*see* **Note 26**).
8. Add 1.25 volumes 100% ethanol to the aqueous phase and mix thoroughly (e.g., if 300 μl was recovered, add 375 μl ethanol) (*see* **Note 27**).
9. For each sample, place a Filter Cartridge into one of the Collection Tubes (supplied in kit).
10. Pipette 700 μl of the lysate/ethanol mixture onto the Filter Cartridge. For sample volumes >700 μl, apply the mixture in successive applications to the same filter.
11. Centrifuge at 10,000 × *g* for ~15 s, or until the mixture has passed through the filter. Alternatively, use a vacuum to pull the samples through the filter.
12. Discard the flow-through and repeat until all of the lysate/ethanol mixture has been passed through the filter. Save the Collection Tube for the washing steps.
13. Add 700 μl miRNA Wash Solution 1 (working solution mixed with ethanol) to the Filter Cartridge.
14. Centrifuge at 10,000 × *g* for ~15 s or use a vacuum to pull the solution through the filter. Discard the flow-through from the Collection Tube and replace the Filter Cartridge into the same Collection Tube.
15. Apply 500 μl Wash Solution 2/3 (working solution mixed with ethanol) and pass it through the filter as in the previous step.
16. Repeat with a second 500 μl of Wash Solution 2/3.
17. After discarding the flow-through from the last wash, replace the Filter Cartridge in the same Collection Tube and centrifuge the assembly at 10,000 × *g* for 1 min to remove residual fluid from the filter.
18. Transfer the Filter Cartridge into a fresh Collection Tube (supplied in kit).
19. Apply 50 μl of preheated (95 °C) Elution Solution or nuclease-free water to the center of the filter.
20. Centrifuge for ~30 s to recover the RNA.
21. Repeat the elution once more with an addition of 50 μl of Elution Solution or nuclease-free water.

22. Collect the elute (which contains the RNA) and place it on ice for immediate use, or store it at ≤ –20 °C.
23. Concentration of RNA from exosome samples is extremely low and is often not detectable when measured using standard spectrophotometers or fluorometers. Instead, the concentration of an RNA solution can be determined by using the Agilent 2100 Bioanalyzer (*see* **Note 28**).

The purified RNA can be analyzed in different ways, for example, retrotranscription to cDNA for expression studies or to build the libraries for sequencing.

4 Notes

1. The cells are maintained in culture until an approximately 70–80 % confluence for adherent cells, and 60–70 % for cells in suspension. The culture medium is removed by washing with PBS two times. From this moment, they are maintained in culture for 24–48 h (depending on the speed of division of cells) with culture medium containing FBS-depleted exosomes. After that time, supernatant samples for purification of exosomes are collected. For cells in suspension, centrifuge 10 min at 300 × *g*, 4 °C, pour off the supernatant, and resuspend cells in the same volume of medium + depleted exosome FBS. Exosomes-depleted FBS is commercial or can be obtained by ultracentrifugation.
2. All biological liquids are extracted in tubes with lithium heparin anticoagulant, except the cerebrospinal fluid to be drawn into tubes without any additive; all tubes with separator carry agar. Blood samples are drawn in tubes with agar and coagulation accelerator.
3. If we start from frozen samples, these are thawed in a bath with water at room temperature (RT) (23–25 °C) until they are completely liquid.
4. The precipitation methods offer advantages over classical methods ultracentrifugation:
   - Reduction in the processing time of the samples (8–30 h for ultracentrifugation is reduced to 2 h for precipitation).
   - The ultracentrifugation limits the number of samples to be processed generally to six (capacity of rotors), whereas the precipitation can process as many samples as possible simultaneously.
   - Less volumes from smaller samples.
5. The antibodies can be used for purification of total exosomes (e.g., tetraspanins CD9 or CD63) or isolated subpopulations

using cell line (e.g., MHC-II antigen-presenting cells or CD3 for T lymphocytes). Thus, depending on what cell type, the markers for detection may vary. As other compartments of the cell also can produce vesicles, it is further recommended to determine the presence of proteins from these compartments, such as the endoplasmic reticulum (e.g., calnexin and Grp78) and the Golgi apparatus (e.g., GM130). Thus, lack of these proteins indicates no or little contamination of vesicles of other compartments in the studied sample.

6. The method of isolating exosomes based on antibody-coated magnetic beads is also available to be performed in 96-well plates.
7. Antibodies can also be adhered to the surface of a ball of latex or sepharose, in this case instead of using a magnetic holder to collect spheres, centrifuge at 10,000 × *g* for 10 min at 4 °C.
8. When removing the supernatant with the pipette, hold the tube at an angle so that the pellet is always covered with supernatant, and stop removing supernatant when half centimeter of liquid is still covering the pellet
9. For this high-speed centrifugation, all tubes should be at least three-quarters full. If one of the tubes is not three-quarters full, add PBS. Centrifugation time is calculated to allow a full hour at 100,000 × *g*, i.e., 10 min for the centrifuge to reach 100,000 × *g* plus 1 h at the final speed. A longer time (up to 3 h) will not damage the exosomes.
10. For fixed-angle rotors, at this step, pour off the supernatant rather than use a pipet. For swinging-bucket rotors, remove the supernatant with a pipette and leave 2 mm of supernatant above the pellet
11. There will probably not be a visible pellet at this step. For fixed-angle rotors, resuspend by flushing up and down where the pellet should be (upper side of the tube, toward the bottom). For swinging-bucket rotors, flush the bottom of the tube
12. If the final volume of exosomes is too large (>1/2000 of the initial volume of conditioned medium) or if there was no visible pellet at **step 10**, use **step 11**.
13. The biological fluids may exhibit a higher degree of viscosity than the culture media; for this reason, it may be advisable to dilute them in 1× PBS at a ratio of ½.
14. As in the cell culture media-derived exosomes, the concentration is smaller, decrease the proportion mixing 1 part/2 parts reagent per sample.
15. To samples of biological fluids, the precipitation efficiency is increased by incubating the sample at 4 °C overnight.

16. After centrifugation, the sample may be filtered through 0.22 μm filters and eliminate the contaminants that remain on the sample. This step is important for serum or plasma, which can drag more pollutants, lipid particles with sizes exceeding 220 nm.
17. Resuspend between ¼ and ½ of the sample volume starting in 1× PBS.
18. The amount of sample to be used in protocols for serum precipitation is in the range of 100–200 μl. This volume can range from 5 μl to 5 ml as a function of the process we make a posteriori.
19. If you need to isolate exosomes in sterile conditions, wash the centrifuge tubes with 70 % ethanol and sterile 1× PBS, leaving them dry in hood.
20. An extra purification of the exosomes can be done by centrifuging 70 min at 100,000 × *g*, 4 °C in a gradient of 30 % sucrose.
21. If we have exosomes resuspended in excess of 200 μl of PBS 1×, an extra centrifugation to concentrate exosomes is needed.
22. Incubation of 5–10 min at RT can promote dissolution of the pellet. Finally, gently pipetting up and down the sample.
23. To dissolve the precipitates, heat the 2× Denaturing Solution at 37 if necessary.
24. For some exosome samples isolated from cell culture media, the volume may be greater than 200 μl. In such cases, take 200 μl for processing, or process the entire volume. When processing >200 μl, keep the 2× Denaturing Solution and the Acid-Phenol:Chloroform proportionate to the starting sample volume.
25. The volume of Acid-Phenol: Chloroform should be equal to the overall volume of the sample plus the 2× Denaturing Solution (e.g., if the initial sample lysate volume was 200 μl and it was mixed with 200 μl of 2× Denaturing Solution in **step 1**, add 400 μl Acid-Phenol: Chloroform).
26. Preheat the Elution Solution or nuclease-free water to 95 °C for eluting the RNA from the filter at the end of the procedure. Nuclease-free water can be used in place of the elution buffer, especially if the RNA is concentrated with a centrifugal vacuum concentrator.
27. 100 % Ethanol must be at room temperature. If the 100 % ethanol is stored cold, warm it to room temperature before starting the RNA isolation.
28. The quality of exosomal RNA cannot be directly analyzed because the exosomal RNA lacks the two prominent 18S and 28S rRNA subunits. Cell and exosome RNAs can be run in parallel and the cellular RNA quality is used as a reference for the exosomal RNA quality.

References

1. Mathivanan S, Ji H Simpson RJ et al (2010) Exosome: extracellular organelles important in intercellular communication. J Proteomics 73(10):1907–1920
2. Lee Y, EL Andaloussi S, Wood MJA (2012) Exosomes and microvesicles: extracellular vesicles for genetic information transfer and gene therapy. Hum Mol Genet 21(R1):R125–R134
3. Blanchard N, Lankar D, Faure F et al (2002) TCR activation of human T cells induces the production of exosomes bearing the TCR/CD3/zeta complex. J Immunol 168:3235–3241
4. Raposo G, Nijman HW, Stoorvoge W et al (1996) B lymphocytes secrete antigen-presenting vesicles. J Exp Med 183(3):1161–1172
5. Thery C, Regnault A, Garin J et al (1999) Molecular characterization of dendritic cell-derived exosomes. Selective accumulation of the heat shock protein hsc73. J Cell Biol 147(3):599–610
6. Wolfers J, Lozier A, Raposo G et al (2001) Tumor-derived exosomes are a source of shared tumor rejection antigens for CTL cross-priming. Nat Med 7(3):297–3037
7. Waldestrom A, Gennebäck N, Hellman U et al (2012) Cardiomyocyte microvesicles contain DNA/RNA and convey biological messages to target cells. PLoS One 7(4):e34653
8. Ronquist G (2012) Proteasomes are mediators of intercellular communication: from basic research to clinical implications. J Intern Med 271(4):400–413
9. Choi DS, Kim DK, Kim YK et al (2014) Proteomics of extracellular vesicles: exosomes and ectosomes. Mass Spectrom Rev 34(4):474–490
10. Caby MPD, Vincendeau-Scherrer C et al (2005) Exosomal-like vesicles are present in human blood plasma. Int Immunol 17(7):879–887
11. Pisitkun T, Shen RF, Knepper MA (2004) Identification and proteomic profiling of exosomes in human urine. Proc Natl Acad Sci U S A 101(36):13368–13373
12. Admyre C, Johansson SM, Qazi KR et al (2007) Exosomes with immune modulatory features are present in human breast milk. J Immunol 179(3):1969–1978
13. Admyre C, Grunewald J, Thyberg J et al (2003) Exosomes with major histocompatibility complex class II and co-stimulatory molecules are present in human BAL fluid. Eur Respir J 22(4):578–583
14. Vlassov AV, Magdaleno S, Setterquist R et al (2012) Exosomes: Current knowledge of their composition, biological functions, and diagnostics and therapeutic potentials. Biochim Biophys Acta 1820(7):940–948
15. Al-Nedawi K, Meehan B, Micallef J et al (2008) Intercellular transfer of the oncogenic receptor EGFRvIII by microvesicles derived from tumour cells. Nat Cell Biol 10(5):619–624
16. Szeles A, Henriksson M et al (2001) Horizontal transfer of oncogenes by uptake of apoptotic bodies. Proc Natl Acad Sci U S A 98(11):6407–6411
17. Holmgren L, Szeles A, Rajnavölgyi E et al (1999) Horizontal transfer of DNA by the uptake of apoptotic bodies. Blood 93:3956–3963
18. Lösche W, Scholz T, Temmler U et al (2004) Platelet-derived microvesicles transfer tissue factor to monocytes but not to neutrophils. Platelets 15(2):109–115
19. Schageman J, Zeringer E, Mu Li et al (2013) The complete exosome workflow solution: from isolation to characterization of RNA cargo. BioMed Res Int 2013:15 p
20. Théry C, Zitvogel L, Amigorena S (2002) Exosome: composition, biogénesis and function. Nat Rev Immunol 2:569–579
21. Janowska-Wieczorek A, Wysoczynski M, Kijowski J et al (2005) Microvesicles derived from activated platelets induce metastasis and angiogenesis in lung cancer. Int J Cancer 113:752–760
22. Lakkaraju A, Rodriguez-Boulan E (2008) Itinerant exosomes: emerging roles in cell and tissue polarity. Trends Cell Biol 18(5):199–209
23. Rajendran L, Honsho M, Zahn TR et al (2006) Alzheimer's disease beta-amyloid peptides are released in association with exosomes. Proc Natl Acad Sci U S A 103:11172–11177
24. Peinado H, Alečković M, Lavotshkin S et al (2012) Melanoma exosomes educate bone marrow progenitor cells toward a pro-metastatic phenotype through MET. Nat Med 18:883–891
25. Admyre C, Telemo E, Almqvist N et al (2008) Exosomes—nanovesicles with possible roles in allergic inflammation. Allergy 63:404–408
26. Suntres EZ, Smith MG, Momen-Heraviv F et al (2013) Therapeutic uses of exosomes. Exosomes Microvesicles 1:2013

27. Valadi H, Ekström K, Bossios A et al (2007) Exosome-mediated transfer of mRNAs and microRNAs is a novel mechanism of genetic exchange between cells. Nat Cell Biol 9:654–659
28. Huang X, Yuan T, Tschannen M et al (2013) Characterization of human plasma-derived exosomal RNAs by deep sequencing. BMC Genomics 14:319
29. Mathivanan S, Simpson RJ (2009) ExoCarta: a compendium of exosomal proteins and RNA. Proteomics 9:4997–5000
30. Kim DK, Kang B, Kim OY et al (2013) EVpedia: an integrated database of high-throughput data for systemic analyses of extracellular vesicles. J Extracell Vesicles 2:20384

Chapter 12

Cell Culture Techniques: Corticosteroid Treatment in A549 Human Lung Epithelial Cell

Elena Marcos-Vadillo and Asunción García-Sánchez

Abstract

Experimentation with cell cultures is widespread in different areas of basic science as well as in the development of biotechnology applied to medicine. Cellular models are applied to the study of respiratory diseases. In this chapter, we present a protocol of basic cell culture using A549 Human lung adenocarcinoma epithelial cells as model. Corticosteroid therapy is used to treat respiratory diseases such asthma. Thus, we also describe a protocol of lung epithelial cell culture treated with dexamethasone to illustrate an example of monitoring the effects of a drug in a lung cell culture assay.

Key words Cell culture, Cell freezing, Cell thawing, Subculture, Corticosteroids, Mycoplasma

1 Introduction

Cell cultures are essential tools not only in basic research but also in the development of the biotechnology industry, without forgetting its relevance in applied medicine [1]. They are used in a wide variety of research areas such as virology, cancer research, or immunology, and allow further studies of intracellular activity, intracellular flow, or cellular interactions [2]. In the study of respiratory diseases such as asthma, it is very common to find reports based on cell cultures of human bronchial epithelial cells [3], human T cells [4], or murine lung fibroblast [5], used to analyze the underlying mechanisms, as well as, the possible development of new therapeutic targets.

There are two different categories of cell culture, the primary culture, and cell line. The first are obtained directly from animal tissues having 100% of the original karyotype. Cell lines are formed, however, by aneuploidy cells mostly from tumoral origin. Primary cultures have a limited life span. They grow a limited number of passages before senescing, whereas cell lines have unlimited proliferation and are maintained during much more passages.

This cellular model of in vitro study confers some advantages over animal experimentation: (a) cell cultures provide a controlled

María Isidoro-García (ed.), *Molecular Genetics of Asthma*, Methods in Molecular Biology, vol. 1434,
DOI 10.1007/978-1-4939-3652-6_12,

environment, allowing a strict control of their physicochemical properties; (b) a large number of homogeneous cells is easily obtained; (c) when possible, the replacement of animal testing by cell cultures may avoid ethical conflicts related to animal experimentation; and (d) cell cultures are more economical than animal models. However, it has some disadvantages such as instability; due to the aneuploidy, the cell lines can generate subpopulations when they are submitted to a large number of passages, and the cell culture cannot always replace the in vivo testing, especially in biomedical research.

Corticosteroid therapy is used to treat a variety of inflammatory and immune disorders, among others, respiratory diseases such asthma. In this chapter, treatment with dexamethasone is used to illustrate an example of monitoring the effects of a drug in a lung cell culture assay.

2 Materials

2.1 Solutions and Reagents

1. 0.4% Trypan Blue Solution, 0.2 μm filtered.
2. 1× Trypsin-EDTA solution: 0.5 g porcine trypsin, 0.2 g EDTA, 4Na per liter of Hanks' Balanced Salt Solution with phenol red.
3. Dimethyl Sulfoxide (DMSO).
4. Dexamethasone: Perform a stock solution at a concentration of 2.5×10^{-3} M in absolute ethanol. Make aliquots and store at −80 °C.
5. Ethanol absolute
6. 70% Ethanol: 70.14 ml of 99.8% ethanol and add water to 100 ml.
7. Sterile Fetal Bovine Serum (FBS).
8. Sterile Modified Fetal Bovine Serum like Charcoal Stripped Serum (Gibco).
9. RPMI 1640 medium with L-Glutamine with Phenol Red and without HEPES.
10. Antibiotics: Penicillin (100 U/ml)/Streptomycin (100 μg/ml). Anti-microbial combination.
11. 1× PBS (Phosphate-Buffered Saline).
12. "RNeasy Plus Mini Kit" (Qiagen).
13. Turbo DNAse-free kit (Ambion).
14. Master Mix Promega (Promega).
15. Oligonucleotides: Mico1 5′-GGC GAA TGG GTG AGT AAC ACG-3′ and Mico2 5′-CGG ATA ACG CTT GCG ACT ATG-3′.

16. Superscript III First Strand Synthesis System for RT-PCR (Life Technologies).
17. LightCycler 480 Sybr Green I Master (Roche).
18. RedSafe Nucleic Acid Staining Solution (iNtRON).
19. Water, PCR grade.
20. Liquid nitrogen.

2.2 Instrumentation and Consumables

1. Laminar flow cabinet.
2. CO_2 incubator.
3. Thermostatic bath.
4. Hemocytometer.
5. Cell culture vessels (Roux culture flasks, Petri dishes, multi-well plates).
6. 1.5 ml microcentrifuge tubes.
7. 15 ml sterile tubes.
8. 50 ml sterile tubes.
9. Cryo vials.
10. Automatic micropipettes and appropriate filter tips.
11. Automatic pipettor.
12. Disposable serological pipettes (5, 10, and 25 ml).
13. Inverted phase contrast microscopy.
14. Centrifuge.
15. Microcentrifuge.
16. Water purification equipment, balances, pH meter.
17. Ice bucket.
18. Sterilization, autoclaves.
19. Spectrophotometer.
20. PCR Thermocycler.
21. qPCR Thermocycler.
22. Electric cold (4, −20, and −80 °C).
23. Cryo preservation module (i.e., StrataCooler Cryo preservation module, Stratagene).

3 Methods

The methods described below are designed for the A549 cell line from human lung epithelial cells. When working with other cell lines, protocols may suffer some modification. Furthermore, volumes of various reagents are specific for the specified container size and must be adapted in each specific case.

In cell cultures, maintaining aseptic measures is essential to avoid possible contamination to ensure cell viability and the veracity of the experimental results [6]:

1. Clean the surface of the laminar flow hood with 70 % ethanol before experimentation.
2. Clean all those tempered containers, in the water bath with 70 % ethanol before using.
3. Avoid touching the surface of the mouths of the media bottles, reagents and cell containers, pipetting with inclined flask.
4. Loose the caps of all reagents and flasks, but do not remove them completely, to facilitate pipetting. Close immediately after use. Keep them open the minimum time possible.
5. Discard any material or culture if you suspect that it can be contaminated.
6. Conduct periodic cleaning of the cabin, incubator, and thermostatic bath to avoid contamination.
7. If the cells are grown in culture flasks, Roux leave plugs slightly loose to ensure an adequate gas exchange.

3.1 Preparing the Subculture (Passage) of (Non-primary) Adherent Cell Line

For primary cultures, the subculture ratio is 1:2. It is essential to estimate the time required to double the cell population. In continuous cell lines, the subculture ratios can be greater and ascertaining the time of "population dubbing" is not so necessary. The standard is a passage of the cells every 2–3 days at a 1: 3 to 1: 5 ratio (*see* **Notes 1** and **2**).

The criteria to determine when it is necessary to perform a subculture are as follows [7]:

- Cell culture density: It is necessary to make a passage before the cells reach the state of confluence.
- Impoverishment of the culture medium.
- Time since last subculture.
- Especial requirements of the experiment, the need to increase the number of cells or to change the type of culture medium.

Although obvious, it is essential to perform a microscopic inspection of the culture before any experimental procedure. Use an inverted phase contrast microscope (100–200×) to quick test the general cell appearance checking that there is no evidence of contamination or dead cells floating in the liquid.

3.1.1 Trypsinization

1. Remove all culture medium of the culture flask Roux 75 cm^2 (T-75 flask) where cells are growing.
2. Wash once with 5 ml of Phosphate-Buffered Saline (PBS) to remove traces of Fetal Bovine Serum (FBS) that can inhibit

trypsin (*see* **Note 3**). Then slowly rock it back and forth to remove all traces of fetal FBS.

3. Remove the PBS carefully.
4. Add 2 ml of trypsin/EDTA to ensure that the entire surface of the vessel is covered (*see* **Note 4**).
5. Rotate the culture flask to ensure even distribution of trypsin.
6. Incubate 5 min at 37 °C in the CO_2 incubator to allow cell detaching.
7. Gently tap the tube on the palm of your hand to disperse the cells and check at the inverted microscope to confirm that the cells are detached from the surface (*see* **Notes 5–7**).
8. Add 4 ml (twice that the used for trypsin) of pre-warmed complete medium (it must contain serum to neutralize the trypsin) (*see* **Note 8**).
9. Collect the trypsinized cell solution and transfer to a sterile 15 ml tube.

3.1.2 Subculture

1. Centrifuge the trypsinized cell solution, 5 min at $\geq 235 \times g$ and remove the supernatant.
2. Resuspend cell pellet in 6 ml of complete medium.
3. Transfer 2 ml of the resuspended cells to a new vial and add 10 ml of complete medium (*see* **Note 9**).
4. Label the flask with the cell line, date, and passage number.
5. Incubate at 37 °C and 5 % CO_2.

3.2 Determination of the Number and Cell Viability

1. Clean the surface of the chamber cell count or hemocytometer with a solution of 70 % ethanol (*see* **Note 10**).
2. Slightly moisten the side of the camera with water or alcohol and press gently sliding the cover to affix it.
3. Detach the A549 cells grown in monolayer by trypsinization (see trypsinization).
4. Centrifuge the cells for 5 min at $\geq 240 \times g$.
5. Resuspend cells in 1 ml of complete medium by gently pipetting several times to avoid the presence of cell aggregates.
6. In an eppendorf, perform a 1/2 dilution of the cells with Trypan Blue Solution adding 10 μl of the suspension cells to 10 μl of colorant (*see* **Notes 11** and **12**).
7. Mix well by pipetting up and down to ensure an even distribution, or give a brief vortex and let stand 5 min before loading the counting chamber.
8. Take 10 μl with automatic micropipette and load it into the chamber Neubauer cell counting. To calculate the total number of cells, takes into account the dilution factor [2] (*see* **Note 13**).

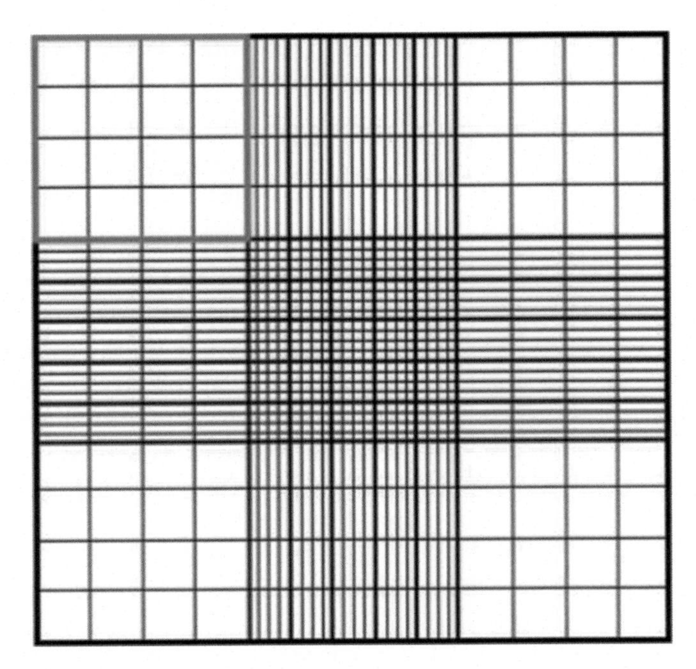

Fig. 1 Diagram of a hemocytometer indicating the 16 corner squares used for counting

9. Make a first check with 10× and 20× objective to center the counting area of the hemocytometer and verify that the cells are distributed evenly across the surface (*see* **Note 14**).
10. Focus on one set of 16 corner squares as indicated in blue in Fig. 1.
11. Switch to the 40× objective and count the number of live cells in this area of 16 squares. Count cells that are within the square and the positioned on the right hand or bottom boundary line (*see* **Note 15**).
12. Determine the concentration of cells per milliliter by the formula:

$$\text{Cells / ml} = \text{average count per square} \times \text{dilution factor} \times 10^4$$

Total cells = cells/ml × total volume of cell suspension from which sample was taken.

13. Determine the percentage of living cells according to formula [8]:

$$\%\ \text{viable cells} = \frac{\text{number of unstained cells}}{\text{total number of cells}} \times 100$$

3.3 Freezing a Cell Line

1. Check possible contamination before freezing.
2. Detach the cells of the culture that you want to freeze (see trypsinization).

3. Centrifuge 8 min at ≥370 × *g*.
4. Carefully decant the supernatant and resuspend the cells in 2 ml of fetal bovine serum.
5. Perform a cell count using a hemocytometer such as a Neubauer chamber or similar, and perform a vital stain with trypan blue (see determining cell number and viability).
6. Adjust to a final concentration of 5–10 × 10^6 cells/ml in fetal calf serum.
7. Put 1.8 ml of the cell solution in a sterile cryo vial and add 200 μl of sterile DMSO (*see* **Note 16**).
8. Freeze immediately at −20 °C for 24 h (*see* **Note 17**).
9. After 24 h, transfer the cryo vials at −80 °C (*see* **Note 18**).
10. After 24 h, move the cells to liquid nitrogen (−170 °C) for long-term storage. Cells are stable at liquid nitrogen for many years.

3.4 Thawing a Cell Line

1. Preheat the medium at 37 °C in a thermostatic bath.
2. Remove the vial from liquid nitrogen tank to dry ice, or from freezer at −80 °C.
3. In a falcon tube, add 30 ml of preheated RPMI in a thermostatic bath.
4. With a sterile Pasteur pipette, add tempered medium to the cell vial and transfer them to the falcon tube (*see* **Note 19**).
5. Centrifuge 5 min at ≥370 × *g*.
6. Meanwhile, prepare a small bottle or plate with 2 ml of complete RPMI, you can put on the stove to be tempered.
7. After centrifuge, discard the supernatant; resuspend the pellet in 3 ml of medium, and transfer to the bottle. Pipette gently up and down to avoid cell agglomerates.
8. For sensitive cells, it is recommended to use medium with 20 % fetal bovine serum and to seed in well plate to be closer.
9. Incubate at 37 °C and 5 % CO_2.
10. Examine with inverted phase contrast microscope 24 h after thawing and subculture if necessary (*see* **Note 20**).

3.5 Pharmacological Treatment with Corticosteroids

Nowadays, one of the most common uses of corticosteroids is the treatment of asthma and other allergic diseases. Thus, inhaled corticosteroids are at the forefront in the treatment of adults and children with persistent asthma [9] or COPD (chronic obstructive pulmonary disease) [10]. In this chapter, treatment with dexamethasone illustrates an example of monitoring the effects of a drug in a lung cell culture assay.

For this test, a modified serum with low concentration in hormones, charcoal stripped serum, is used to supplement the culture medium. This will reduce the potential interference with corticosteroid treatment that could derive from the presence of steroids hormones in a conventional fetal bovine serum.

3.5.1 Cell Culture Treatment

1. Make a stock dilution of dexamethasone to obtain an intermediate concentration of 2.5×10^{-5} M (dilution 1/100) in complete RPMI medium supplemented with 10 % charcoal stripped serum and appropriate antibiotics (*see* **Note 21**).
2. Prepare a 1/100 dilution of absolute ethanol, to treat the cells that serve as basal condition in the experiment (*see* **Note 22**).
3. Seed the A549 cells at a density of 500,000 cells/well in 6-well plates, in a final volume of 1800 μl. Perform three conditions (each in triplicate):
   - Treatment with dexamethasone to a final concentration of 2.5×10^{-6} M.
   - Control of the vehicle (ethanol).
   - Control treatment (cells alone, without any additives).
4. Add 200 μl of 2.5×10^{-5} M dexamethasone to the corresponding treatment wells (the final concentration is 2.5×10^{-6} M), 200 μl of 1/100 ethanol to the vehicle control wells, and 200 μl of RPMI (10% FBS + antibiotics) to the control wells (*see* Table 1).

3.5.2 Collection of Cells

In this protocol, we collect the cells at 36 h.

1. Collect the medium in an eppendorf and store at −80 °C for further determinations.
2. Gently wash the attached cells with 500 μl of PBS.
3. Add 500 μl of trypsin and store in an incubator 5–6 min until cells are detached.
4. Neutralize the trypsin with 1000 μl of complete RPMI medium and transfer the cell suspension to an eppendorf.

Table 1
Design of cell culture for dexamethasone assay

	Dexa treatment	Ethanol control	Treatment control (cells alone)
Cell volume (500000/well)	1800 μl/well	1800 μl/well	1800 μl/well
Complete medium	–	–	200 μl/well
Dexa 2.5×10^{-5}	200 μl/well	–	–
Ethanol 1/100	–	200 μl/well	–
Total volume	2000 μl/well	2000 μl/well	2000 μl/well

5. Centrifuge at ≥370 × g 5 min.
6. Remove the supernatant and wash the pellet with 500 µl of PBS.
7. Centrifuge at ≥370 × g 5 min.
8. Remove the supernatant and add 350 µl of the lysis buffer of RNA extraction kit.
9. Vigorously mix the tube on the vortex to disrupt the cells and put away immediately at −80 °C until RNA extraction.

3.5.3 RNA Extraction

Keep the RNA samples on ice to prevent RNA degradation. Clean the work surface and pipettes used (preferably pipettes designed only to work with RNA) with 70 % ethanol. Make the work as quickly as possible to avoid the RNA degradation.

1. Add the lysate directly to a QIA shredder column and centrifuge 2 min to V max to homogenize.
2. Add 350 µl (1 vol) of 70 % ethanol to homogenize the lysate and mix well by pipetting. Do not spin.
3. Pipette 700 µl of the sample to a QIAamp new column. Centrifuge 15 s to ≥8000 × g. If you had more volume, repeat the centrifugation with the rest.
4. Add 500 µl of Buffer RPE to the column and centrifuge at ≥8000 × g for 15 s to wash.
5. Add 500 µl of Buffer RPE and centrifuge at ≥8000 × g for 3 min.
6. Place the column into a new tube and centrifuge at full speed for 1 min.
7. Transfer the column to a 1.5 ml tube and pipette 30 µl H_2O RNAse free directly on the membrane. Centrifuge at ≥8000 × g for 1 min to elute the sample.
8. Measure the concentration in Nanodrop. Verify that the 260/280 nm absorbance ratio, which indicates the purity of the extracted RNA, is around 1.8 accepting that the RNA is of good quality. Furthermore, the 260/230 ratio must be between 2.0 and 2.2 considering that ratios below this value show a phenolic solutions contamination that may alter subsequent experimental procedures [11].

3.5.4 RNA Treatment with DNase

1. Transfer the 30 µl of RNA to a PCR tubes and add:
   - 2.9 µl of 10× DNase Buffer.
   - 1 µl Turbo DNase.
2. Make a brief spin.
3. Incubate 30 min, 37 °C.
4. Make a brief spin.
5. Add 3 µl of inactivation resin. Mix well the resin before pipetting it.

6. Incubate the mixture 5 min at room temperature (25 °C). Mix occasionally.
7. Centrifuge 2 min, ≥8000 × *g*.
8. Collect the supernatant careful not to catch the resin and transfer to a new 1.5 ml tube.
9. Store the RNA at −20 °C if it is to be used soon, or −80 °C for long periods.

3.5.5 RNA Retrotranscription

1. For the retrotranscription, use 1 μg of RNA considering the concentration measured before DNase treatment. Prepare the following RNA-primer mixture:
   - 1 μg of RNA.
   - 1 μl of 510 ng/μl random hexamers.
   - 1 μl of 10 mM dNTP mix.
   - DEPC-treated water to 10 μl.
2. Incubate the tube at 65 °C for 5 min, then place on ice 1 min.
3. Prepare the following cDNA synthesis mix reaction adding the components:
   - 2 μl of 10× RT buffer.
   - 4 μl of 25 mM $MgCl_2$.
   - 2 μl of 0.1 M DTT.
   - 1 μl of RNase out 40 U/μl.
   - 1 μl of SuperScript III RT 200 U/μl.
4. Add 10 μl of cDNA synthesis mix reaction to each RNA-primer mixture, mix gently, and collect by brief centrifugation (*see* **Note 23**).
5. Program the thermocycler as follow:
   - 25 °C—10 min.
   - 50 °C—50 min.
   - 85 °C—5 min.
6. Incubate on ice.
7. Spin for collect the reactions and add 1 μl of RNase H to each tube and incubate at 37 °C—20 min.

   The cDNA is obtained in a final volume of 20 μl and an estimated concentration of 50 ng cDNA/μl [12].

3.5.6 Quantitative PCR (qPCR)

Work in a laminar flow cabinet.

1. For each sample, 20 ng of cDNA are used, prepare the following reactions:
   - Three reactions with specific oligonucleotides of gene to study.

- Three reactions with oligonucleotides of a housekeeping gene (GAPDH in our experiment).

2. Preparation of samples:
   - 3.2 μl of cDNA + 36.8 μl of H_2O. Add to each well 5 μl of this sample mixture.
3. Preparation of the reaction mixtures:
   - 1.3 μl H2O.
   - 7.5 μl LightCycler 480 SybrGreen I Master.
   - 1.2 μl Primer F/R 10 μM.
   - Multiply by the number of reactions that will use the same oligos and make a mix for all.
4. Preparation of the plate: Add to each well 5 μl of the cDNA mixture and 10 μl of the corresponding reaction mixture.

3.6 PCR Assessment of Mycoplasma Contamination

The main source of Mycoplasma contamination in a cell culture is the use of animal serum or trypsin contaminated. Certain mycoplasma species are found in human skin, and a poor aseptic technique can contaminate the cultures. It is a contamination of small prokaryotic cells that can form colonies, which may go unnoticed since they are not as obvious as contamination by bacteria or yeast. The Mycoplasma contamination can cause different effects on culture cells, altering metabolism and proliferation. Therefore, it is necessary to conduct regular checks of cell culture [2].

Commercial kits have been developed to detect the presence of Mycoplasma species responsible for the majority of cell culture contamination by detecting 16S RNA of 8 mycoplasma species (*M. hyorhinis, M. arginini, M. pneumoniae, M. fermentans, M. orale, M. pirum, Acholeplasma laidlawii, and Spiroplasma mirum*) (*see* **Note 24**).

1. For best results, maintain culture cells in the absence of antibiotics for several days in order to observe a good signal in the PCR.
2. Take a 100 μl of culture supernatant. This can be stored at 4 °C for a couple of days before to be used.
3. Boil or heat to 95 °C the supernatant for 5 min. This step breaks the mycoplasma membrane and releases the DNA.
4. Spin the sample briefly in the microcentrifuge. Transfer the supernatant to a new tube and freeze at −20 °C for futures PCR analyses.
5. Keep on ice while preparing PCR.
6. Prepare PCR reaction:
   - 1 μl each Primer Mico1/Mico2 (10 pmol/μl each).
   - 12.5 μl Master Mix.

- 8.5 μl RNAse-free Water, PCR grade.
- 2.0 μl Sample.
- The total PCR volume will be 25 μl.

7. Program the thermocycler:

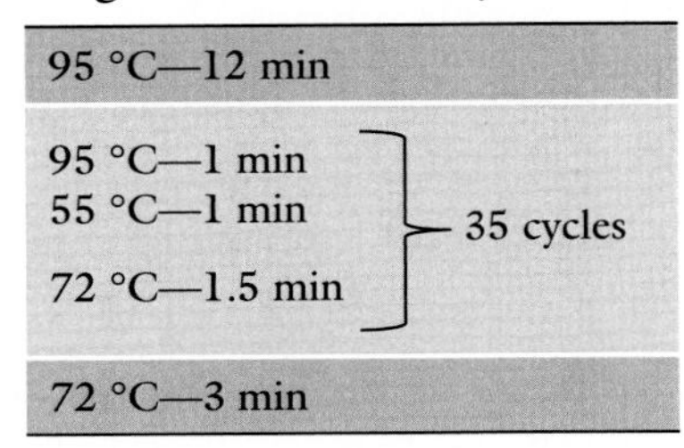

8. While the PCR is in progress, prepare the 1 % agarose gel in TBE (Tris-Borate-EDTA):
   - 0.5 g agarose.
   - 50 ml 0.5 % TBE buffer.
   - 1.5 μl RedSafe Nucleic Acid Staining Solution.
9. Once the PCR is finished, and the agarose gel is in the electrophoresis chamber, load the PCR sample on the gel and perform the electrophoresis.
10. In the case of a positive sample for mycoplasma, a fragment of 500 bp will appear.

4 Notes

1. Both the passage time and the cells subculture ratio may vary between cell lines or in different conditions.
   Change the ratios and the time monitoring the cell growth in the inverted microscope preventing a complete cell confluence, 100 % confluence may alter the properties of the cell line. In addition, the more confluent the cells, the more difficult to disperse into a single cell suspension.
2. In slow growing cell cultures, change the medium every 2 or 3 days by replacing 2/3 of the old medium for a new tempered medium.
3. Adjust volumes according to the bottle surface used (Table 2) [13].
4. In cultures where cells are well lifted, trypsin can be diluted to the medium with serum-free medium or PBS.
5. Adherent cells have a rounded morphology when not attached to any surface.
6. If we check at inverted microscope that some of the cells continue attached on the surface, increase incubation at 37 °C

Table 2
Volume of medium according to the culture devices

	Description	Surface (cm^2)	Volume (ml)
Flasks	T-25	25	5–10
	T-75	75	15–25
	T-150	150	30–50
Cell culture dishes	35	8	1–2
	60	21	4–5
	100	55	10–12
Multiwell plates	24-Well	1.88	0.5–1.2
	12-Well	3.83	1.0–2.4
	6-Well	9.40	2.0–3.0

1 or 2 min; be careful not to expose the cells for long time as the trypsin action can damage them.

7. If the cells are 100 % confluent, trypsinization is more difficult because cell-cell interactions prevent the enzyme to reach the interface between the cell and the substrate.
8. Add a double volume of medium with serum and then the trypsin volume to counteract the trypsin.
9. Cell lines should be subcultured in order to prevent culture dying. Transfer only the appropriate amount of cells for the required cell density (see the ECACC data sheet for the cell line). For A549, split sub-confluent cultures (70–80 %) 1:3 to 1:6, i.e., seeding at $2–4 \times 10^4$ cells/cm^2.
10. The surface of the count chamber must be clean and free of dust and watermarks.
11. Living cells look healthy and appear colorless when microscopically observed, while dead cells are blue stained.
12. Pipetting gently up and down to catch cells evenly.
13. Dispense the dilution slowly and check that is positioned between the coverslip and the slide.
14. If the cells are not evenly distributed or cell aggregates are observed, clean the camera and reload after pipetting the initial solution.
15. It is recommended that the density should be between 20 and 50 cells per 1×1 mm square, for counting on camera. If not, perform the appropriate dilution, or concentrate the cells by centrifugation homogenizing in a smaller volume of complete medium.

16. Depending on the requirements of the cell line, the cells can be frozen in DMEM with 20% serum or 100% serum, in any case with the presence of 10% DMSO as cryoprotectant.
17. It is advisable to transfer the cell to a special rack, i.e., StrataCooler Cryo preservation module, designed to freeze mammalian cells at a controlled rate of 0.4–0.6 °C/min to achieve an 80–90% survival rate. In this case, transfer the cryo vials to the prechilled (4 °C) StrataCooler Cryo preservation module and place it in −80 °C freezer.
18. The cells can be kept frozen at −80 °C for 1–2 months.
19. Alternatively you can also thaw the vial containing the cells in a 37 °C bath. It is important to thaw quickly to minimize any cell damage. Transfer the cells to the falcon tube with sterile Pasteur pipette. Remember to clean the vial with a tissue soaked in 70% ethanol before opening in the hood, to avoid contamination. However the other method is preferred.
20. Adjust the volume of the medium and the flask size to achieve the correct seeding density for the cell line.
21. The vial should always be kept at −80 °C. At the time of use, remove from freezer; make an intermediate dilution with culture medium, returning it to the freezer to −80 °C as quickly as possible. Always protect from light. The intermediate dilution should not be reused.
22. An assay with the basal condition should be done in parallel to any assay for studying the effect of a treatment. To this, cells should be treated following the same procedure but using the vehicle in which the reagent is dissolved, ethanol in this case.
23. It is advisable to scale up the reaction mixture as needed.
24. There are alternative methods for the detection of mycoplasma, such as ELISA.

Acknowledgments

This work was supported by grants of the Junta de Castilla y León ref. GRS1047/A/14 and BIO/SA73/15; and by the project "Efecto del Ácido Retinóico en la enfermedad alérgica. Estudio transcripcional y su traslación a la clínica," PI13/00564, integrated into the "Plan Estatal de I+D+I 2013–2016" and cofunded by the "ISCIII-Subdirección General de Evaluación y Fomento de la investigación" and the European Regional Development Fund (FEDER).

References

1. Mirón A (2011) Riesgo biológico: evaluación y prevención en trabajos con cultivos celulares. Noticias Técnicas de Prevención 902 Instituto Nacional de Seguridad e Higiene en el Trabajo
2. Cultivos celulares. Documento de aplicación. CULTEK. http://www.cultek.com/inf/otros/soluciones/Cultivos%20Celulares/Aplica_Cultivos_Celulares_2007.pdf. Accessed 13 Sept 2014
3. Tanabe T, Shimokawaji T, Kanoh S et al (2014) Secretory phospholipases A2 are secreted from ciliated cells and increase mucin and eicosanoid secretion from goblet cells. Chest. doi:10.1378/chest.14-0258
4. Jackson DJ, Makrinioti H, Rana BM et al (2014) IL-33-dependent type 2 inflammation during rhinovirus-induced asthma exacerbations in vivo. Am J Respir Crit Care Med 190:1373–1382
5. Wongtrakool C, Grooms K, Bijli KM et al (2014) Nicotine stimulates nerve growth factor in lung fibroblasts through an NFkB-dependent mechanism. PLoS One 9(10):e109602. doi:10.1371/journal.pone.0109602
6. Freshney RI (2010) Culture of animal cells: a manual of basic technique, 6th edn. Wiley-Blackwell, New York, pp 57–71
7. Culture of Animal Cells. Basic Techniques (2012) http://lifescience.roche.com/wcsstore/RASCatalogAssetStore/Articles/Culture%20_of_Animal_Cells-Basic_Techniques_TT.pdf. Accessed 20 Sept 2014
8. Phelan MC (1998) Basic techniques for mammalian cell tissue culture. Curr Protoc Cell Biol 1.1.1–1.1.10. John Wiley & Sons, Inc
9. Barnes PJ (2006) Corticosteroids: the drugs to beat. Eur J Pharmacol 533:2–14
10. Barnes PJ (2011) Glucocorticosteroids: current and future directions. Br J Pharmacol 163:29–43
11. 260/280 and 260/230 Ratios. Technical bulletin. Thermo Scientific. http://www.nanodrop.com/Library/T009-NanoDrop%201000-&-NanoDrop%208000-Nucleic-Acid-Purity-Ratios.pdf. Accessed 17 Sept 2014
12. User Bulletin #2: ABI PRISM 7700 Sequence Detection System (2011) Appl Biosyst
13. (2014) Animal cell culture guide: tips and techniques for continuous cell lines. ATCC® http://www.atcc.org. Accessed 5 Dec 2014

Chapter 13

Protocol for Lipid-Mediated Transient Transfection in A549 Epithelial Lung Cell Line

Elena Marcos-Vadillo and Asunción García-Sánchez

Abstract

Trials of transfection in eukaryotic cells are essential tools for the study of gene and protein function. They have been used in a wide range of research fields. In this chapter, a method of transient transfection of the A549 cell line, human lung cells of alveolar epithelium, with an expression plasmid is described. In addition, the fundamental characteristics of this experimental procedure are addressed.

Key words A549, Expression vector, Lipofectamine, Lipofection, Transfection

1 Introduction

1.1 Cellular Transfection

In the field of molecular and cellular genetics, the term transfection refers to the process whereby exogenous genetic material is introduced into cultured eukaryotic cells using nonviral mechanisms. The development of techniques taking as base model the introduction of DNA in *E. coli* (bacterial transformation) has notoriously driven the understanding of gene regulation, protein expression, and cellular function [1].

The transfection techniques allow performing protein expression in a eukaryotic system in which posttranslational modifications necessary for correct operation (absent in prokaryotes models) are carried out. Two types of transfection can be distinguished referring as transient or stable transfection, respectively [2]. Transient transfection provides a high degree of gene expression and allow co-express several proteins at the same time. But, only temporal protein expression, with high variability in the percentage of expressing cells will be obtained.

Conversely, when a stable transfection is carried out, the resulting cell population will be a homogeneous clonal population with theoretically continuous production of protein. In this case, the level of expression will be moderate and it will be difficult to

María Isidoro-García (ed.), *Molecular Genetics of Asthma*, Methods in Molecular Biology, vol. 1434,
DOI 10.1007/978-1-4939-3652-6_13, © Springer Science+Business Media New York 2016

simultaneously co-express multiple proteins. In addition, the selection of the clones that have integrated the plasmid is tedious and requires the incorporation of a plasmid containing a gene for antibiotic resistance to select and perpetuate the clones [3].

1.2 Transfection Methods

For an efficient transfection, different barriers such as the entry of genetic material into the cell, release of the endosome, escape of lysosomal degradation, and translocation to the nucleus need to be overcome. Several transfection methods have been developed in recent decades to solve these problems [4].

Since DNA molecule and cell membranes are both negatively charged, the electrostatic opposing forces have to be neutralized to surmount the first barrier. To this end, numerous methods have been developed. Generally, transfection assays can be divided into chemical methods, seeking to coat the DNA molecules with positive charges, and physical methods, whereby pores are created in cell membrane or the DNA molecule is mechanically introduced into the cell [5–11].

An ideal transfection method would meet the following criteria: it should be easy to perform, with minimum costs and efforts; the method should be suitable for both stable and transient transfection; and finally the procedure should be highly versatile, i.e., it should be valid for any cell type [12] (*see* Table 1).

1.3 Expression Vectors

The progress in the understanding of mammalian genes has been largely due to the use of expression vectors in cell culture. Many proteins are expressed at low levels in cells culture and therefore expression vectors that increase protein synthesis by strong promoters are used. In addition, the expression vectors allow characterizing the impact of specific mutations on cell metabolism and their ability to stably alter the cellular phenotype as a function of transgene expression [14].

Depending on the final purpose of the study, the expression system that best fits the needs should be chosen. There are vectors that amplify the synthesis of fusion proteins using a strong promoter along with the genes of interest. Other systems allow studying the mechanisms of gene expression regulation by combining promoters and transcription factors to the gene of interest, or to analyze the effects caused by certain treatments on the protein expression.

The expression vectors should facilitate their use in this type of assay and contribute to efficiently express the exogenous gene within the target cell [15]. Although the characteristics may vary depending on the study peculiarities, they can be summarized as follows:

- A prokaryote autonomous replication origin that permits plasmid amplification in bacterial systems.
- A eukaryotic replication origin that makes possible the plasmid expression in the target cells.

Table 1
Characteristics of different transfection methods

Method		Details	Comments: Advantages	Comments: Disadvantages
Chemicals	Calcium phosphate	DNA molecules precipitate in calcium phosphate; an insoluble complex is introduced into the cell by endocytosis	Simplicity Stable and transient transfection	Low reproducibility Variable efficiency In vitro techniques only
	DEAE-dextran	Cationic polymer binding to DNA is introduced into the cell by endocytosis	Simplicity High efficiency Reproducibility Low cost	Transient transfection Toxicity In vitro techniques only
	Lipofection	Formation of colloidal particles with surrounding lipid membranes DNA molecules are introduced into the cell by endocytosis or membrane fusion	High efficiency Applications in vitro and in vivo Not immunogenic High versatility Friendly Large DNA molecules	High cost More effective in adherent cells than in cells suspension
Physical	Electroporation	Open pores in the plasma membrane by electric pulse of high intensity and short duration	Simplicity High reproducibility High efficiency Transient and stable transfection	High cell death rates Requires cell resuspension steps High cost
	Microinjection	Introduction of macromolecules through fine glass capillaries of solutions under control by microscopy	High efficiency High versatility	Complex technique Optimization of multiple parameters Handling of individual cells (high time consumption) High cost
	Biolistic	Microparticle bombardment. Genetic material bond to biologically inert particles (tungsten or gold) penetrates the cell membrane at high speed	Useful in hard-to-transfect cells Useful in immunization trials [13] High reproducibility Low amount of DNA Little cell manipulation	Lower efficiency electroporation or lipofection Preparation of micro particles High cost Requires specific instrumentation

- A promoter of the gene of interest recognized by a mammalian polymerase. The signal sequences of the transcription start are essential for foreign protein expression. Most common promoters are viral, although other bacteriophage promoters such as constitutive or tissue specific or even inducible promoters can also be used.
- A multiple cloning region where the DNA encoding the protein will be inserted.
- A transcription terminator.
- A ribosome-binding site.
- Antibiotic resistance genes that allow a first selection of the bacterial clone with the plasmid, when the vector amplification is performed in a bacterial system. The most widely used is the resistance to ampicillin. Other selection genes can be incorporated when a stable transfection is carried out, for example, neomycin resistance gene or dihydrofolate reductase, which confers resistance to methotrexate.
- Regulatory gene sequences.
- Fragment of DNA for homologous recombination in order to achieve stable transfections.

An ideal vector should [16]:

- Be reproducible.
- Be stable.
- Permit the insertion of genetic material without size restriction.
- Reach high concentrations.
- Enable specific integration of gene.
- Discriminate and act on specific cells.
- Be under control/regulation.
- Be fully characterized.
- Be innocuous and with negligible or no side effects.
- Be easy to produce and store at a reasonable cost.

The huge number of published papers that employ this technology corroborates the wide range of applications resulting from the use of expression vectors. Some of the most remarkable applications are:

- The study of recombinant proteins.
- Therapeutic applications of modified cells.
- Studies of protein expression and protein function.
- Studies of gene regulation (promoters, regulatory elements, etc.) [17].
- Generation of transgenic organisms.

- Gene Therapy.
- The expression of protein for purification.
- The study of RNA processing.
- The study of protein interaction.
- Subcellular localization of proteins.

In this chapter, a method of transient transfection of the A549 cell line, human lung cells of alveolar epithelium, will be described. The cells will be transfected with a sequence fragment of the prostaglandin D2 receptor *PTGDR*, a gene involved in asthma development.

2 Materials

2.1 Reagents

2.1.1 Culture

1. Human alveolar basal epithelial cells, A549: They can be purchased in American Type Culture Collection (ATCC) website (http://www.atcc.org).
2. Trypan Blue Solution 0.2 μm filtered.
3. 1× Trypsin-EDTA solution.
4. 70 % Ethanol.
5. Fetal Bovine Serum (FBS) sterile.
6. RPMI 1640 medium with L-Glutamine (or other appropriate medium).
7. Antibiotics: Penicillin (100 U/ml)/Streptomycin (100 μg/ml). Anti-microbial combination.
8. 1× Phosphate-Buffered Saline (PBS): 1.37 mM NaCl, 2.6 mM KCl, 10 mM Na_2HPO_4, 1.8 mM KH_2PO_4. Adjust the pH to 7.4 with HCl.
9. H_2O sterile.

2.1.2 Transfection

1. Opti-MEM® I Reduced Serum Medium (Gibco).
2. Lipofectamine® 2000 Reagent (Invitrogen).
3. pCMV6-Entry plasmid: Expression vector with strong constitutive CMV promoter.
4. pCMV6-PTGDR plasmid: pCMV6-Entry vector in which the gene encoding the PTGDR has been cloned in the polylinker.
5. pUC18 plasmid: Vector used as DNA carrier, to stabilize the total amount of DNA transfected in all experimental conditions.

2.2 Instrumentation and Consumables

1. 24-Well cell culture cluster (other culture plates can be used).
2. Laminar flow cabinet.
3. CO_2 incubator.

4. Thermostatic bath.
5. Hemocytometer.
6. 1.5 ml microcentrifuge tubes.
7. 15 ml sterile tubes.
8. 50 ml sterile tubes.
9. Automatic pipettes and appropriate tips.
10. Electric pipettor.
11. Inverted phase contrast microscopy.
12. Centrifuge.
13. Microcentrifuge.
14. NanoDrop 1000 spectrophotometer (Thermo Fisher Scientific).
15. Commercial purification kits plasmid DNA (i.e., Qiagen Plasmid Maxi kit).

3 Methods

The lipofection methods are not universal, and vary widely among different cell lines. The optimal conditions should be identified for each cell type and for each type of test. Before any testing with cell lines, we recommend consulting the American Type Culture Collection (ATCC) website (http://www.atcc.org) for helpful information about optimal growth media and specific serum.

It is essential to maintain aseptic measures to avoid possible contamination ensuring cell viability and veracity of results. The whole process should be done in laminar flow cabinets, decontaminate all surface with 70% ethanol, use sterile material and discard any material, or culture if they are suspected to be contaminated (*see* Fig. 1 for quick view protocol)

The type of plasmid DNA vector is an important factor. In this experiment, the vectors used are pCMV6-Empty, pCMV6-PTGDR, and pUC18 (*see* Fig. 2).

Plasmids are amplified by bacterial transformation, purified using commercial purification kits and eluted in H_2O. Before starting the transfection process, prepare the vectors at the appropriate concentration of use. The concentration is spectrophotometrically quantified. Verify that the 260/280 nm absorbance ratio, which indicates the purity of the extracted DNA, is around 1.8. Furthermore, the 260/230 ratio must be between 2.0 and 2.2; ratios below this value indicate phenolic contamination that may alter subsequent procedures [18]. Dilutions are recommended to obtain a working solution. Plasmid solutions are aliquoted and stored at −20 °C until use.

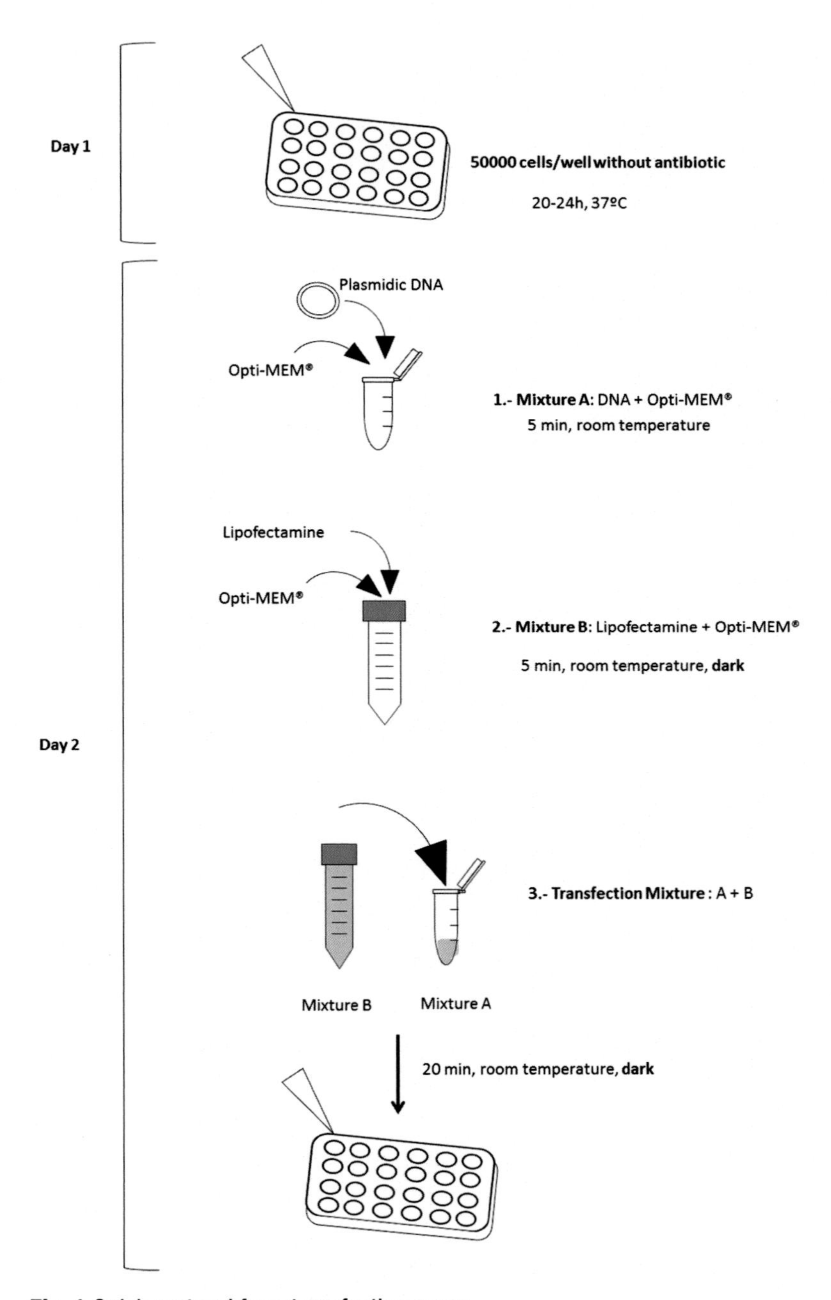

Fig. 1 Quick protocol for a transfection assay

3.1 Seed Cells (See Notes 1 and 2)

1. Remove and discard the culture medium of the T-75 flask where cells are growing.
2. Wash with 5 ml of PBS to remove all serum traces that contains trypsin inhibitor. Then, carefully remove the PBS.

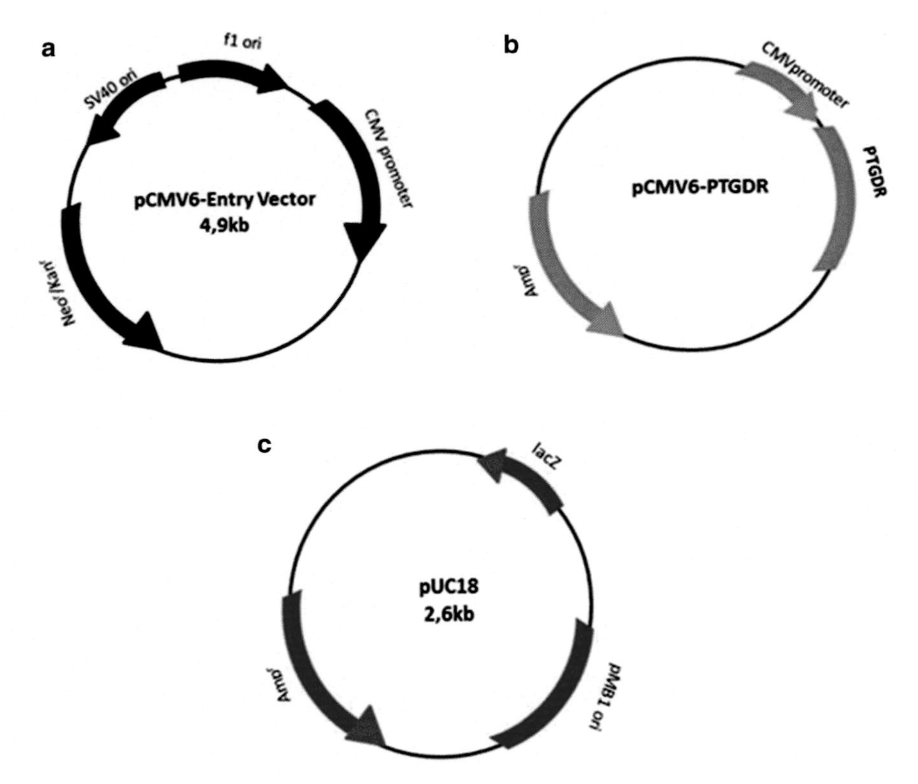

Fig. 2 Expression vector maps

3. Detach adherent cells by trypsinization. Add 2 ml of trypsin/EDTA and incubate 5 min at 37 °C in the CO_2 incubator to allow cell detaching. Check at the inverted microscope to confirm the cells are detached from the surface.
4. Neutralize the trypsin with 4 ml of pre-warmed complete RPMI medium by incubating in a 37 °C water bath (*see* **Note 3**).
5. Recount the cells by vital staining with Trypan Blue solution, counting in Neubauer chamber or similar.
6. Resuspend the cells in a volume of pre-warmed medium to obtain a density of 1×10^6 cells/ml.
7. Seed 50,000 cells per well in a total volume of 500 μl of RPMI medium supplemented with 10% fetal bovine serum without antibiotics (*see* **Notes 4–6**).
8. Ensure uniform distribution of the cells in the wells and prevent cell clump (*see* **Note** 7).
9. Maintain 20–24 h in incubation at 37 °C and 5% CO_2. The cell density should be between 60 and 80% at the time of performing the transfection process (*see* **Notes 8** and **9**).

It is required to make appropriate controls in each transfection assay. Include a negative control without DNA to check optimal cell growth conditions. To analyze problems related with the insert,

Table 2
Example of a transfection assay

Volumes (μl)	Single cells	25 ng empty vector	25 ng expression vector	50 ng empty vector	50 ng expression vector	100 ng empty vector	100 ng expression vector
pCMV6-PTGDR (25 ng/μl)	–	–	1	–	2	–	4
pCMV6 empty (25 ng/μl)	–	1	–	2	–	4	–
pUC18 (carrier) 250 ng/μl (up to 500 ng)	2	1.9	1.9	1.8	1.8	1.6	1.6
Optimem (up to 50 μl)	48	47.1	47.1	46.2	46.2	44.4	44.4

transfect the cells with the plasmid without the gene of interest (i.e., include only the plasmid backbone used to construct the expression vector). Including a positive control is also advisable.

3.2 Transfection (See Note 10)

3.2.1 Mixture A

1. Dilute the plasmid DNA in Opti-MEM I Reduced Serum Medium in the corresponding pre-labeled Eppendorf tubes.
2. According to the manufacturer instructions, transfect cells with 500 ng of final DNA per well, in a final volume of 50 μl of Opti-MEM I Reduced Serum Medium per well. Perform each reaction mix in triplicate.(*see* **Notes 11–14**) (*see* Table 2).
3. Incubate at least 5 min at room temperature.

3.2.2 Mixture B

1. Prepare 1 μl of lipofectamine and 49 μl of Opti-MEM I Reduced Serum Medium per well. Mix by pipetting up and down several times (do not vortex) (*see* **Note 15**).
   - *Lipofectamine: 1 μl/well×3.5 (triplicate)×n° tubes mixture A.*
   - *Opti-MEM: 49 μl/well×3.5 (triplicate)×n° tubes mixture A.*
2. Incubate 5 min at room temperature and protected from light.

3.2.3 Transfection (Mixture A+B)

1. Add 175 μl of mixture **B** (lipofectamine + Opti-MEM) to each tube of 1.5 ml containing mixture **A** (DNA + Opti-MEM), (Ratio 1:1) (*see* **Note 16**).
2. Incubate 20–30 min at room temperature to allow formation of the DNA-liposome complex. Protect from light (*see* **Note 17**).

3.2.4 Incubation of Cells with the Transfection Mixture

1. Use cells seeded the day before. Remove the culture medium and perform washing with PBS carefully not to disturb the cells (*see* **Notes 18** and **19**).
2. Add to each well 100 μl of the transfection mixture **A + B** (DNA+ lipofectamine + Opti-MEM I Reduced Serum Medium).

3. Add to each well 100 μl of de Opti-MEM I Reduced Serum Medium to ensure that the cells are completely covered with medium.
4. Maintain the cells in the CO_2 incubator with the transfection mixture for 4–6 h.
5. After the incubation period, remove transfection mixture, with care not to lift the cells, and add 200 μl complete RPMI medium with antibiotics. Optional: perform washes with PBS between media change.
6. Maintain 12–72 h in incubation at 37 °C and 5 % CO_2 (*see* **Note 20**).

3.3 Harvest Cells to Analyze Transfection

The collection of the cellular extracts can be performed from 12 h after transfection up to 72 h [14], depending on cell type and the activity of the promoter integrated the vector.

Collect the culture medium in an eppendorf or in a cryo vial and store at −80 °C for further determinations.

1. Gently wash the attached cells with 200 μl of PBS per well.
2. Add 200 μl of trypsin/EDTA per well, and incubate 5–6 min at 37 °C in the CO_2 incubator until cells are detached.
3. Neutralize the trypsin with 400 μl of complete RPMI medium per well, and transfer the cell suspension to an eppendorf.
4. Centrifuge at $\geq 370 \times g$ 5 min.
5. Remove the supernatant and wash the pellet with 500 μl of PBS.
6. Centrifuge at $\geq 370 \times g$ 5 min and remove the supernatant.
7. Freeze the cell pellet at −80 °C or resuspend it in appropriate buffer for further determinations.

4 Notes

1. Cells must be in exponential growth phase at the time of transfection. Do not use cells that are in or above 30 passages after thawing the vial of a stock culture, to prevent changes in the characteristics of the cell culture as a result of repeated passages [19].
2. The volumes of the transfection reaction are scalable depending on the tissue culture. It is recommended to perform triplicates for each condition of the experiment, in order to ensure the reproducibility of the assay.
3. In order to avoid possible contamination, after tempering or heating any vial in a water bath, wash the surface of the container with 70 % ethanol before introducing it in the laminar flow cabinets.

4. Fetal bovine serum (FBS) can reduce the efficiency of transfection. It contains heterogeneous concentrations of hormones, proteins, and other biomolecules that may interfere in subsequent experimental results. Evaluate the possibility of eliminating or reducing the concentration of fetal bovine serum from 10% to 2–3%, at the time of seeding cells for expression assays. Notice that absence of serum can make the liposome to be more toxic, and the reduction of FBS levels slows cell growth and may reduce cell viability causing low transfection efficiencies [20]. An alternative to serum starvation is using a modified serum with low concentrations of some of its components
5. Under general conditions, the medium for cell culture is supplemented with antibiotics to prevent contaminations. However, lipofectamine, or cationic lipids cause an increase in the permeability of cellular membranes. As a result, the antibiotic penetrates into the cells in greater quantities causing cytotoxicity and deterioration of the transfection efficiency [19].
6. Change cell numbers proportionally for different size plates.
7. For optimal results, it is important to have a single cell suspension. It is advisable to pipette up and down when seeding the cells, and gently rock the plate to ensure good distribution.
8. It is recommended to maintain the cell culture at a density of 60–80% of confluence, to ensure high efficiency in the transfection process (*see* Fig. 3).

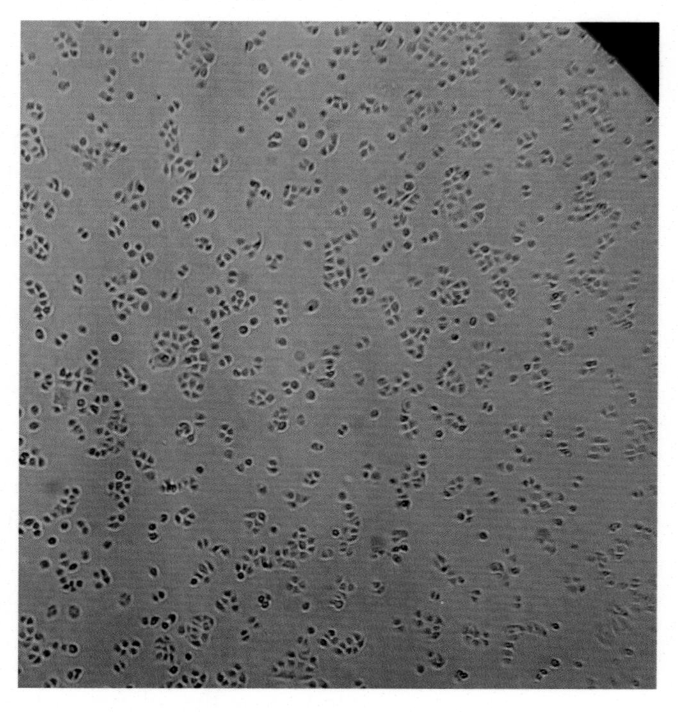

Fig. 3 Cell density for transfection assays

Adjust the number of seeded cells and transfect when cells are 60–80 % confluent.

9. In order to stabilize and recover the culture from trypsinization, it is recommended to carry out the cell seeding at least 24 h prior to transfection.
10. Although there are currently commercial reagents not interfered by fetal bovine serum, the presence of serum in the culture medium can inhibit the transfection by interfering the DNA-lipid complexes. In such cases, it is recommended to dilute the DNA and reagents of transfection in specific medium with low serum concentration such as Opti-MEM® I Reduced Serum Medium.
11. The promoter coupled to the expressed gene must be compatible with the cellular system chosen. Size and topology (linear or supercoiled) of the plasmid DNA vector affect the transfection efficiency.
12. Use an innocuous plasmid vector to complete up to 500 ng per well when it is not possible to use the target vector (i.e. pUC18).
13. Follow the manufacturer's specifications for the amount of DNA to transfect. The optimal amount of DNA used in transfection assays need to be optimized depending on cell type, experimental conditions, and the procedure used. The transfected DNA molecule must be free of protein, RNA, and chemical contaminants such as formaldehyde, isopropanol, or ethanol used in the purification process.
14. Check the accurate quantity of expression vector to transfect (*see* Table 2).
15. It is recommended to perform a whole reaction mix for triplicates taking into account pipetting variations, in order to save time and reagents (i.e., multiply the single reaction volume by 3.5 to get enough transfection mix for 3 replicate wells including an extra for pipetting error).
16. The DNA: lipofectamine ratio can be varied from 1: 0.5 to 1: 5 to adjust the efficiency.
17. Incubations that last more than 30 min may result in a decrease in transfection efficiency.
18. Ensure that cells are covered with solution to prevent drying and suffering. Accordingly, the exchange of the culture medium, washing, and change of transfection medium must be quickly performed.
19. When washing, deposit the PBS on the wall, by gently pipetting, not directly on the cells.
20. The time of incubation with the transfection mixture can vary depending on the cell type used.

Acknowledgments

This work was supported by grants of the Junta de Castilla y León ref. GRS1047/A/14 and BIO/SA73/15; and by the project "Efecto del Ácido Retinóico en la enfermedad alérgica. Estudio transcripcional y su traslación a la clínica," PI13/00564, integrated into the "Plan Estatal de I+D+I 2013–2016" and cofunded by the "ISCIII-Subdirección General de Evaluación y Fomento de la investigación" and the European Regional Development Fund (FEDER).

References

1. Jiménez-Sánchez A, Guerrero R (1982) Genética molecular bacteriana. Reverté (ed) I.S.B.N: 84-291-5544-9
2. Lodish H (2005) Biología celular y molecular. Médica Panamericana (ed) I.S.B.N: 9500613743
3. Walker MR, Rapley R (1997) Strategies for identifying desirable recombinant clones. In: Route maps in gene technology, Blackwell Publishing Ltd., Oxford, UK. doi: 10.1002/9781444313611.ch72
4. Rocha A, Ruiz S, Coll JM (2002) Estudios sobre translocación de complejos DNA-liposomas a través de membranas celulares para mejorar la transfección en células eucariontes (revisión). Invest Agr Prod Sanid Anim 17:5–20
5. Felgner PL, Gadek TR, Holm M et al (1987) Lipofection: a highly efficient, lipid-mediated DNA-transfection procedure. Proc Natl Acad Sci U S A 84:7413–7417
6. Kim TK, Eberwine JH (2010) Mammalian cell transfection: the present and the future. Anal Bioanal Chem 397:3173–3178
7. Technical Reference Guide. An introduction to transfection methods. http://bio.lonza.com/uploads/tx_mwaxmarketingmaterial/Lonza_BenchGuides_An_Introduction_to_Transfection_Methods__Technical_Reference_Guide.pdf Accessed 10 Jan 2015
8. Transfection Methods Overview. http://www.bio-rad.com/webroot/web/pdf/lsr/literature/10-0826_transfection_tutorial_interactive.pdf. Accessed 20 Dec 2014
9. Maurisse R, De Semir D, Emamekhoo H et al (2010) Comparative transfection of DNA into primary and transformed mammalian cells from different lineages. BMC Biotechnol 10:9
10. Khan KH (2010) Gene transfer technologies and their applications: roles in human diseases. Asian J Exp Biol Sci 1:208–218
11. Ma B, Zhang S, Jiang H et al (2007) Lipoplex morphologies and their influences on transfection efficiency in gene delivery. J Control Release 123:184–194
12. Castro FO, Portelles Y (1997) Transfección de ADN a células de mamíferos. Biotecnol Apl 14:149–161
13. O'Brien J, Lummis SCR (2011) Nano-biolistics: a method of biolistic transfection of cells and tissues using a gene gun with novel nanometer-sized projectiles. BMC Biotechnol 11:66
14. Colosimo A, Goncz KK, Holmes AR et al (2000) Transfer and expression of foreign genes in mammalian cells. Biotechniques 29: 314–331
15. Kaufman RJ (2000) Overview of vector design for mammalian gene expression. Mol Biotechnol 16:151–160
16. Khan K (2013) Gene expression in mammalian cells and its applications. Adv Pharmaceut Bull 3:257–263
17. Manetsh M, Ramsay EE, King EM et al (2012) Corticosteroids and b2-agonists upregulate mitogen-activated protein kinase phosphatase 1: *in vitro* mechanisms. Br J Pharmacol 166: 2049–2059
18. 260/280 and 260/230 Ratios. Technical bulletin. Thermo Sci. http://www.nanodrop.com/Library/T009-NanoDrop%201000-&-NanoDrop%208000-Nucleic-Acid-Purity-Ratios.pdf. Accessed 17 Sept 2014
19. http://www.lifetechnologies.com/es/en/home/references/gibco-cell-culture-basics/transfection-basics/factors-influencing-transfection-efficiency.html
20. Protocols and Applications Guide. Transfection guide. Promega. www.promega.com. Accessed 7 Nov 2014

Chapter 14

Promoter Assay Using Luciferase Reporter Gene in the A549 Cell Line

Elena Marcos-Vadillo and Asunción García-Sánchez

Abstract

The development of reporters systems has simplified the study of promoter activity in different areas of knowledge, and represents an easy and fast approach to study genetic variations. In this chapter, we show a transfection protocol of A549 lung epithelial cells with a reporter vector, using the Luciferase-Renilla dual system for studying the variations caused by several polymorphisms in the promoter region of a gene.

Key words Luciferase, Promoter activity, Renilla, Reporter, Transfection

1 Introduction

The use of reporter genes in transfection assays is an essential key element in the study of expression and function gene, enabling a rapid assessment of the transfected DNA effects [1, 2]. One of their major applications is monitoring the transcriptional activity.

Gene reporter is attached to a promoter or regulatory sequence (i.e., specific response elements) that provides an easily measurable signal upon modulation of its expression when introduced into the cell. Thus, the effects of promoter mutations on protein expression can be checked. The characterization and identification of minimal functional sequence of gene promoters involved in development of asthma have been reported, as ORMDL3 [3] and IL-13 [4], among others.

Functional studies to check the association of different single-nucleotide polymorphisms (SNPs) with pathologies, i.e., severe asthma in childhood [5–7], are of special interest. This technique also permits to analyze the effects of different treatments on gene expression and their mechanism of action [8, 9].

Other applications of reporter gene systems are the optimization of the parameters of transfection assays, often complex in its implementation [10], the detection of the precise localization of a protein in the cell, analysis of protein-protein interaction, studies of mediators in trans as factors of transcription, mRNA processing, etc.

María Isidoro-García (ed.), *Molecular Genetics of Asthma*, Methods in Molecular Biology, vol. 1434,
DOI 10.1007/978-1-4939-3652-6_14, © Springer Science+Business Media New York 2016

The reporter genes, therefore, enable to control the efficiency of input DNA into the cell, study the regulation of the expression of a gene, and monitor the subcellular localization of transfected proteins [11].

An ideal system would be a reporter gene that satisfied the following characteristics:

1. Not to endogenously express in the cell.
2. Easily measurable and quantifiable.
3. Reproducible with high sensitivity.
4. Easy handling and equipment.
5. Not supposed to be a high economic effort for the laboratory.

The basis of all reporter system is an expression vector that does not have more regulatory sequences than those introduced, to avoid distortions in the results. In addition, the presence of two multiple cloning sites (MCS) is desirable, one located upstream of the reporter gene, where the promoter sequences are inserted, and another located elsewhere in the vector to incorporate regulatory elements that act at distance [12].

A wide range of reporter genes to monitor gene expression in mammalian cells, including chloramphenicol acetyl transferase (CAT), firefly luciferase, human growth hormone, alkaline phosphatase (AP), and β-galactosidase, have been developed (*see* Table 1) [11–13].

Table 1
Characteristics of different reporter genes

Reporter gene	Detection method	Advantages	Disadvantages
Chloramphenicol acetyltransferase (CAT)	Isotopic detection Fluorescence ELISA	Widely used Stable protein	Radioactivity Low sensitivity Not viable in vivo
Green fluorescent protein (GFP)	Fluorescence (UV exposure)	In vitro and in vivo In situ in tissue samples No need substrate High average life No toxicity [26]	Semiquantitative Not high sensitivity High fluorophore generation time
Firefly luciferase (luc)	Bioluminescence	High specificity and sensitivity Easy, fast High linearity	Expensive equipment No viable cell Protein with short half-life
Alkaline phosphatase (AP)	Colorimetric Chemiluminescence	From transfected cells Cell lysate is not required Assay not disturbed by transfected cells [27]	Endogenous alkaline phosphatase activity
β-Galactosidase (lacZ gene from *E. coli*)	Colorimetric Fluorescence Chemiluminescence Immunohistochemistry	In vitro and in vivo Easily parsed in situ in tissue samples	High background Different sensitivity according to the method Potentially toxic [28]

Firefly luciferase (luc), a monomeric protein of 61 kDa, is one of the most used reporter systems due to its high sensitivity, specificity, easy handling, and the fact that it does not require posttranslational modifications so it is activated [14]. A great advance in the use of the firefly luciferase reporter gene has been the association with the Renilla luciferase gene in co-transfections. The combination of cell transfection with luciferase's vector together with another reporter vector, as Renilla, makes possible to standardize the test by reducing inter- and intra-assay variations such as the number of cells seeded, the efficiency of the transfection, or failures in the measurement procedure. Thus, the Renilla luciferase expression linearly correlates with the expression of firefly luciferase, behaving as an internal control of each experiment.

Firefly luciferase assay is based on a chemiluminescent oxidation-reduction reaction wherein the enzyme substrate (luciferin) combined with ATP, oxygen, and Mg^{2+} results in an oxidized complex, the oxyluciferin, emitting a measurable light. The Renilla luciferase catalyzes a similar reaction in which the oxidation of the substrate, coelenterazine, generates light emission in different wavelength than the firefly luciferase (*see* Fig. 1). The differences in the bioluminescence produced by both enzymes as well as in the substrates make possible discrimination between both enzymatic activities and the joint interpretation [15].

In recent years, many trials have been performed using this dual-reporter vector system in the study of asthma pathology and its treatment [16–19] as well as those designed to elucidate the role played by the miRNA in the development of asthma [20–23].

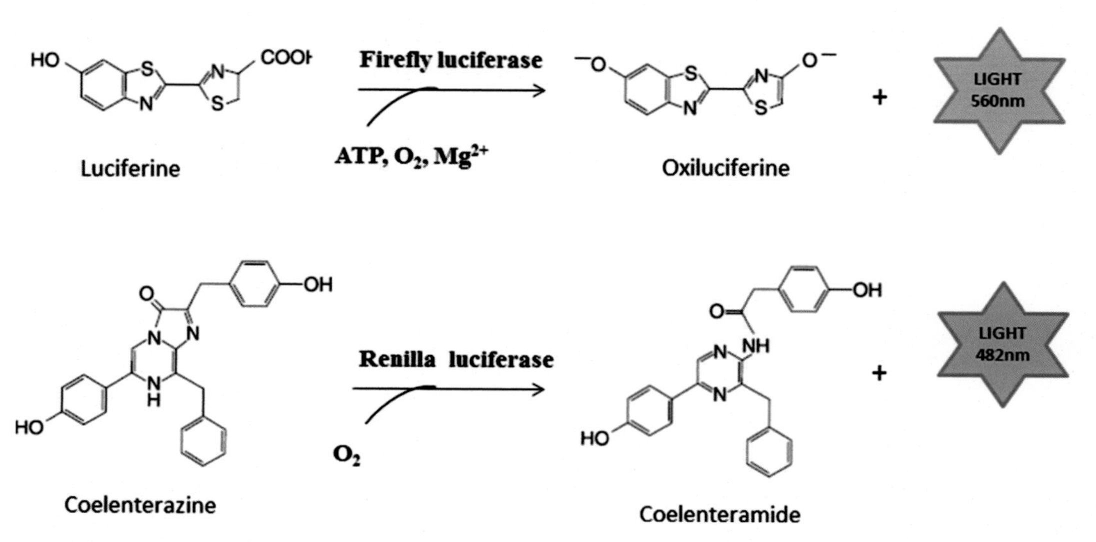

Fig. 1 Luciferase and Renilla chemical reactions

2 Materials

2.1 Reagents

2.1.1 Culture

1. Human alveolar basal epithelial cells, A549: They can be purchased in American Type Culture Collection (ATCC) website (http://www.atcc.org).
2. Trypan blue solution 0.2 μm filtered.
3. 1× Trypsin-EDTA solution.
4. 70% Ethanol
5. Fetal bovine serum (FBS) sterile.
6. RPMI 1640 medium with L-glutamine (or other appropriate medium).
7. RPMI 1640 medium with L-glutamine supplemented with antibiotics and 10% FBS.
8. Antibiotics: Penicillin (100 U/ml)/streptomycin (100 μg/ml).
9. 1× PBS: 1.37 mM NaCl, 2.6 mM KCl, 10 mM Na_2HPO_4, 1.8 mM KH_2PO_4. Adjust the pH to 7.4 with HCl.

2.1.2 Transfection

1. Opti-MEM®I Reduced Serum Medium (Gibco).
2. Lipofectamine® 2000 Reagent (Invitrogen).
3. pGL3-Basic: A vector carrying the firefly luciferase reporter gene inserted. Lacks eukaryotic promoter sequences and other regulatory elements as enhancers. In this way, the expression of luciferase will be only regulated by regulatory sequences introduced by the researcher.
4. pGL3-Basic + *PTGDR*: The gene promoter of interest (in this case *PTGDR*) cloned in the pGL3-Basic vector, upstream of the luc + gene.
5. pRL-SV40: A vector containing the SLV40 enhancer that provides a high level of expression of cDNA coding the Renilla luciferase (Rluc).
6. pUC18: A high-copy-number small plasmid used as carrier DNA, to stabilize the amount of DNA transfected in all experimental conditions.
7. Lysis Passive Buffer (Dual Luciferase System, Promega).
8. Luciferase Assay Reagent II (LAR II): Luciferase Assay Substrate, and Luciferase Assay Buffer II (Dual Luciferase System, Promega).
9. Stop & Glo Reagent: Stop & Glo Buffer and 50× Stop & Glo Substrate (Dual Luciferase System, Promega).

2.2 Instrumentation and Consumables

1. 24-Well cell culture cluster (or other appropriate culture plates).
2. Laminar flow cabinets.

3. CO_2 incubator.
4. Thermostatic bath.
5. Hemocytometer.
6. 1.5 ml Microcentrifuge tubes.
7. 15 ml Sterile tubes.
8. 50 ml Sterile tubes.
9. Automatic micropipettes and appropriate tips.
10. Automatic pipettor.
11. Inverted phase-contrast microscope.
12. Centrifuge.
13. Microcentrifuge.
14. Platform shaker.
15. Luminescence microplate reader instrument.

3 Methods

It is essential to maintain aseptic measures to avoid possible contamination ensuring cell viability and the veracity of the experimental results. The whole process should be done in laminar flow cabinets, decontaminate all surfaces with 70% ethanol, use sterile material, and discard any material or culture if they are suspected to be contaminated.

Before starting the process of transfection, prepare the vectors to be used. In this experiment the vectors used are pGL3-Basic, pGL3-Basic + PTGDR, pRL-SV40, and pUC18 (*see* Fig. 2). Plasmids are amplified by bacterial transformation and purified using commercial DNA plasmid purification kits (i.e., Qiagen Plasmid Maxi kit) and eluted in H_2O.

The plasmid DNA concentration is quantified by spectrophotometry. Verify that the 260/280 nm absorbance ratio is around 1.8. Furthermore, the 260/230 ratio must be between 2.0 and 2.2; ratios below this value indicate a phenolic contamination that may alter subsequent experimental procedures [24]. Plasmid solutions are aliquoted and stored at −20 °C until use.

3.1 Seed Cells (See Note 1) (See Fig. 3)

1. Remove and discard the culture medium.
2. Wash with PBS to remove all traces of serum that contains trypsin inhibitor.
3. Detach adherent cells by trypsinization.
4. Neutralize the trypsin with complete RPMI medium prewarmed by incubating in a 37 °C water bath (*see* **Note 2**).

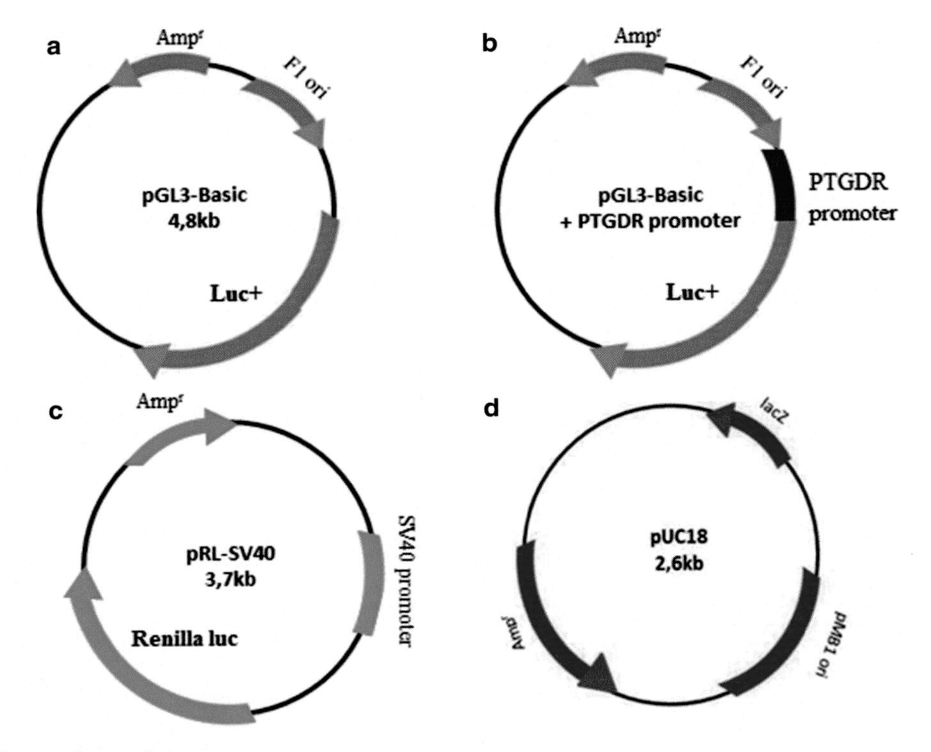

Fig. 2 Expression vector maps

5. Count the starting cells by vital staining with trypan blue solution in a Neubauer chamber or similar.
6. Resuspend the cells in a volume of pre-warmed medium to obtain a concentration of 1×10^6 cells/ml.
7. Seed 50,000 cells per well in a total volume of 500 μl of antibiotic-free RPMI medium supplemented with 10% FBS (*see* **Notes 3–5**).
8. Ensure uniform distribution of the cells per well and prevent cell clumps (*see* **Note 6**).
9. Keep the cells for 20–24 h at 37 °C and 5% CO_2. Cell density should be between 40 and 80% at the time of performing the transfection (*see* **Notes 7** and **8**).

3.2 Transfection

3.2.1 Mixture A

1. Dilute the plasmid with Opti-MEM I Reduced Serum Medium in the corresponding pre-labeled 1.5 ml tubes.
2. According to the manufacturer's instructions transfect cells with 500 ng of final DNA per well, in a final volume of 50 μl Opti-MEM I Reduced Serum Medium per well. Perform each reaction mix in triplicate (*see* **Notes 9–11**) (*see* Table 2).
3. Incubate for at least 5 min at room temperature.

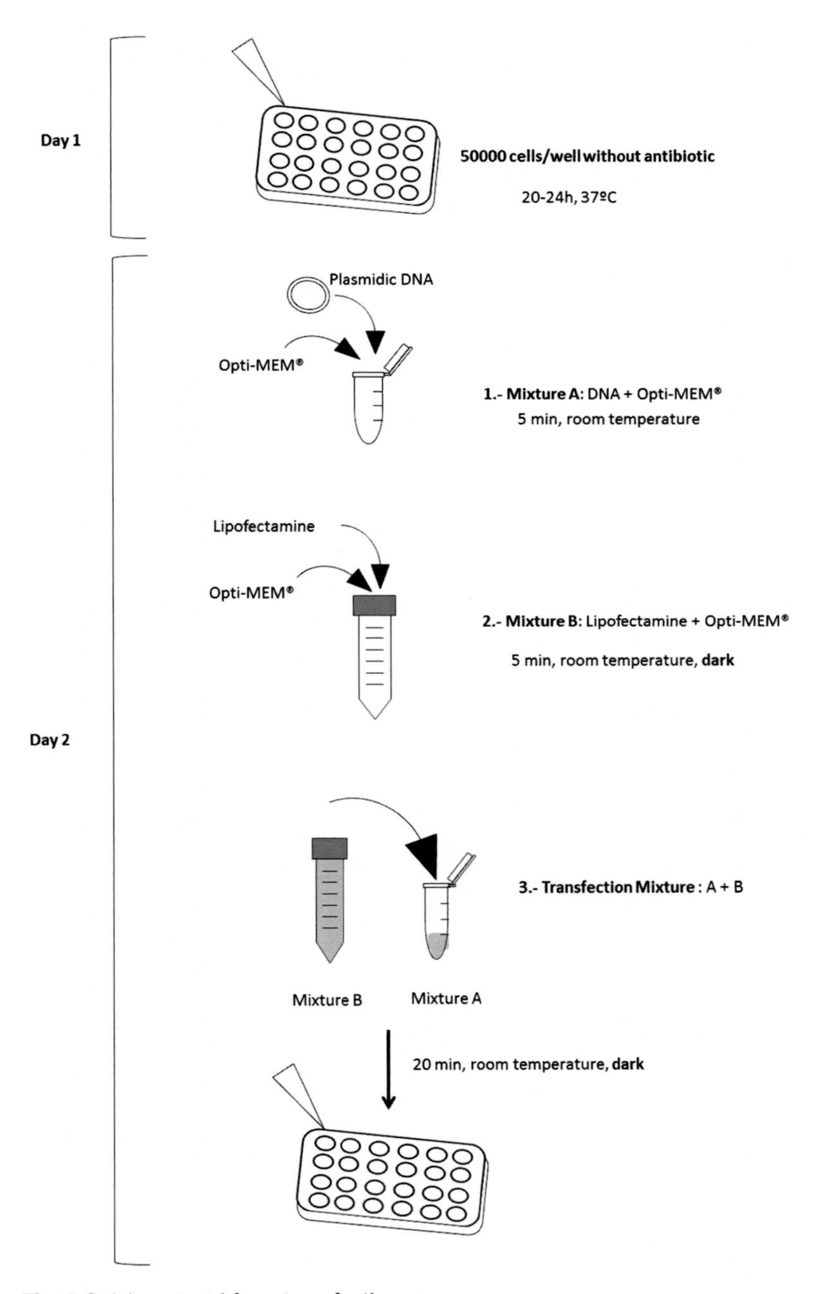

Fig. 3 Quick protocol for a transfection assay

3.2.2 Mixture B

1. Prepare 1 μl of Lipofectamine and 49 μl of Opti-MEM I Reduced Serum Medium per well. Mix by pipetting up and down several times (do not vortex) (*see* **Note 12**).

   - *Lipofectamine: 1 μl/well × 3.5 (triplicate) × n° tubes mixture A.*
   - *Opti-MEM: 49 μl/well × 3.5 (triplicate) × n° tubes mixture A.*

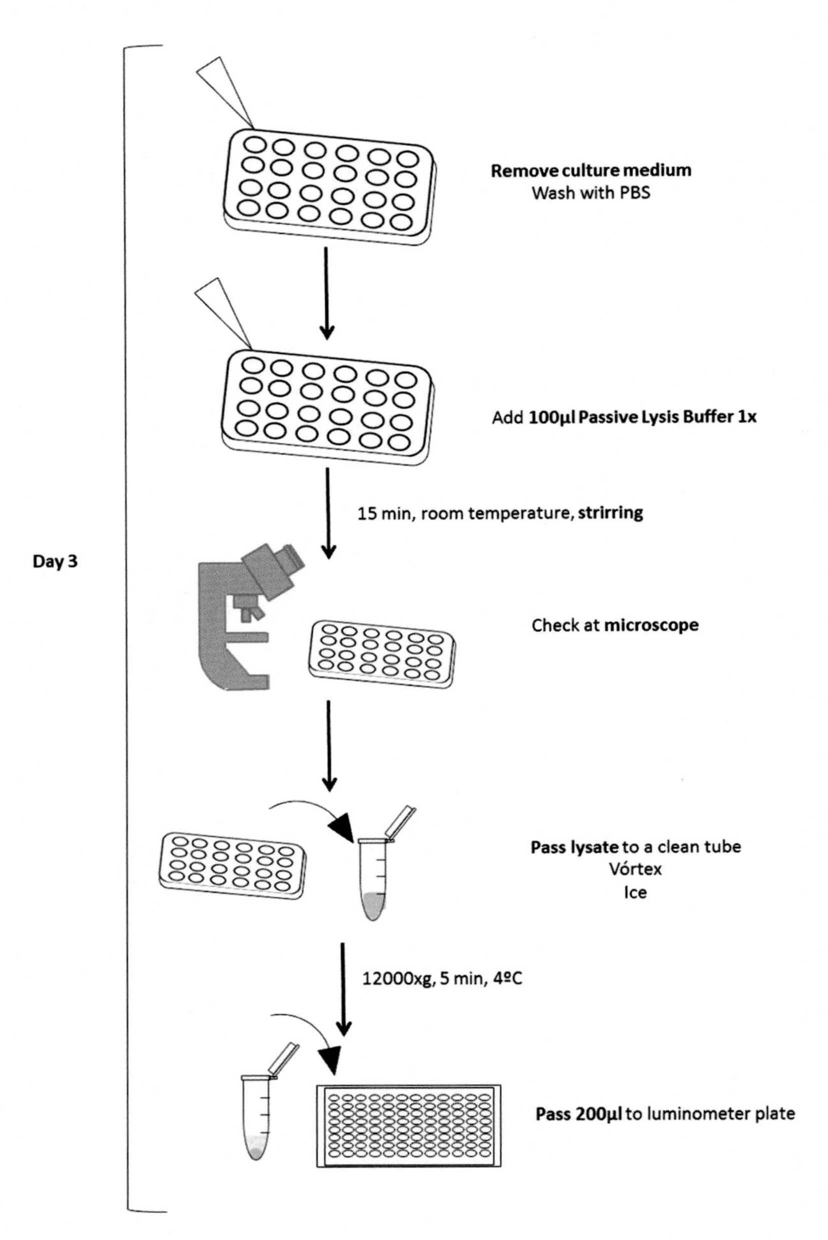

Fig. 3 (continued)

2. Incubate for 5 min at room temperature and protect from light.

3.2.3 Transfection (Mixture A + B)

1. Add 175 µl of mixture **B** (Lipofectamine + Opti-MEM) to each tube containing mixture **A** (DNA + Opti-MEM) (ratio 1:1) (*see* **Note 13**).
2. Incubate for 20–30 min at room temperature to allow formation of the DNA-liposome complex. Protect from light (*see* **Note 14**).

Table 2
Example of an experiment with four polymorphic variants of the target promoter, in 24-well plate

Volumes (μl)[a]	Promotor 1	Promotor 2	Promotor 3	Promotor 4
pGL3Basic + promotor1 (250 ng/μl)	7	–	–	–
pGL3Basic + promotor2 (250 ng/μl)	–	7	–	–
pGL3Basic + promotor3 (250 ng/μl)	–	–	7	–
pGL3Basic + promotor4 (250 ng/μl)	–	–	–	7
pSLV40 (Renilla) (10 ng/μl)	3.5	3.5	3.5	3.5
Opti-MEM (until 50 μl)	164.5	164.5	164.5	164.5

[a]Each condition is calculated in triplicate plus an excess (3.5×)

3.2.4 Incubation of the Cells with Transfection Mixture

1. Remove the cell plates of the CO_2 incubator. Discard the culture medium and perform PBS washes carefully; do not disturb the cells (*see* **Notes 15** and **16**).
2. Add 100 μl of the transfection mixture **A + B** (DNA + Lipofectamine + Opti-MEM) to each well.
3. Add 100 μl of de Opti-MEM I Reduced Serum Medium to each well to ensure that the cells are completely covered with medium.
4. Keep the cells in the CO_2 incubator with the transfection mixture for 4–6 h.
5. After the incubation period, remove the transfection mixture (do not lift the cells). Add 200 μl of complete RPMI medium with antibiotics. Optional: Perform washes with PBS between media change.
6. Keep the cells for 12–72 h in incubation at 37 °C and 5% (*see* **Note 17**).

3.3 Harvest the Cells

1. After 12–72 h remove the cells out of the CO_2 incubator (*see* **Note 18**).
2. Dilute one volume of 5× Passive Lysis Buffer with four volumes of distilled water, and mix well to obtain a final concentration of 1×. Let room temperature before use. Calculate 100 μl of 1× passive lysis buffer per well. Scale up the volumes according to the wells in the experiment. Prepare immediately before use (*see* **Notes 19** and **20**).
3. Remove the culture medium and perform washes with PBS pre-warmed in a water bath.
4. Add 100 μl of 1× Passive Lysis Buffer to each well.
5. Incubate the plate for 15 min at room temperature with shaking.

6. After the incubation check at the microscope that cells are completely lysed. Increase incubation time if needed.
7. Recover the full cell lysate to a new pre-labeled 1.5 ml tube (*see* **Note 21**).
8. Vortex the microcentrifuge tubes for 10–15 s.
9. Keep samples on ice while the reagents are prepared for luciferase signal quantitation or store the tubes at −80 °C until use (*see* **Note 22**).

3.4 Luciferase Assay

1. Calculate a sufficient amount of luciferase and Renilla reagents. Let at room temperature before use. Gently mix by inverting the vial several times before using any reagent (*see* **Note 23**).
   - LAR II (luciferase reagent): Previously prepare the reagent resuspending the lyophilized Luciferase Assay Substrate in 10 ml of Luciferase Assay Buffer II, aliquot, and store at −80 °C. Use 50 μl per well. Thaw in bath at room temperature since that contains reagents that may be altered by heat. Stay protected from the light.
   - Stop & Glo Reagent (Renilla reagent): Prepare immediately before use. Prepare 50 μl per well by diluting the 50× Stop & Glo Substrate with Stop & Glo Buffer to obtain a final concentration of 1×.
2. Centrifuge the lysates at 12,000 × *g* for 5 min at 4 °C.
3. Transfer 20 μl of each lysate supernatant to each well of the 96-well plate for luminometer. (It is possible to perform the analysis with other reading luminometer tube.)
4. Place the plate in the luminometer, and program the instrument such that the injectors 1 and 2 dispense 50 μl of LAR II and Stop & Glo Reagent, respectively.
5. Measure the light produced in the reaction must be 1 or 2 s after the addition of the substrate, with 5–10 s of measurement.

4 Notes

1. It is recommended to perform triplets for each condition of the experiment in order to ensure the reproducibility of the assay.
2. After tempering or heating any vial in a water bath, wash the surface of the container with ethanol 70 % before introducing it in the laminar flow cabinets, in order to avoid possible contamination.
3. Serum can reduce the efficiency of transfection, although the absence of serum can make the liposome more toxic for cells.

4. The liposome complex can interact with the antibiotic decreasing effectiveness and increasing the toxicity.
5. Scale up cell numbers proportionally for different size plates.
6. For optimal results it is important to have a single-cell suspension. To ensure that the cells are distributed homogeneously in the well, gently pipette up and down when seeding the cells.
7. Adjust the number of seeded cells so that at the time of transfection they are 40–80 % confluence.
8. In order to allow the culture to stabilize and recover from the trypsinization, it is recommended to carry out the cell seeding for at least 24 h prior to transfection assay.
9. Use an innocuous vector to complete up to 500 ng per well vector if with the target vector it is not possible (i.e., pUC18).
10. Follow the manufacturer's specifications for the amount of DNA to transfect.
11. Preliminarily, it is suggested to transfect with different amounts of the expression vector in order to choose the most appropriate.
12. It is recommended to perform a whole reaction mix for triplicates and take into account pipetting variations in order to save time and reagents (*see* Table 2).
13. DNA:Lipofectamine ratio can be varied from 1:0.5 to 1:5 to adjust the transfection efficiency.
14. Incubation longer than 30 min may result in a decrease in transfection efficiency.
15. Ensure that the cells are covered with solution to prevent drying and suffering. Accordingly change of the culture medium, washing, and change of transfection medium must be quick.
16. When washing, deposit the PBS by the wall, by gently pipetting.
17. The incubation time with the transfection mixture can vary depending on the cell line used.
18. Culture cell collection for subsequent analyses (DNA, RNA, proteins, cytokines) can be performed from 12 h after transfection up to 72 h [25], depending on cell line and the activity of the promoter integrated the vector.
19. Passive Lysis Buffer is very viscous. Pipette slowly making sure that appropriate volume is caught.
20. Other lysis buffers can increase the luminescence background; if another is used, carry out tests to control this.
21. It is possible to make an active cell lysis by scraping.
22. If the quantification cannot be performed immediately, the lysates can be frozen at −80 °C. In this case it is recommended

to analyze data in the first thaw because several freeze-thaw cycles may result in a decrease in luciferase signal.

23. Follow the manufacturer's instructions when reconstituting reagents.

Acknowledgments

This work was supported by grants of the Junta de Castilla y León ref. GRS1047/A/14 and BIO/SA73/15, and by the project "Efecto del Ácido Retinóico en la enfermedad alérgica. Estudio transcripcional y su traslación a la clínica," PI13/00564, integrated into the "Plan Estatal de I+D+I 2013–2016" and cofunded by the "ISCIII-Subdirección General de Evaluación y Fomento de la investigación" and the European Regional Development Fund (FEDER).

References

1. Ramos E, Espinosa E, Vega A et al (2005) Expresión de genes reporteros en la línea QT 35 utilizando la transfección con liposomas. Rev Salud Anim 27:152–158
2. Jiang T, Xing B, Rao J (2008) Recent developments of biological reporter technology for detecting gene expression. Biotechnol Genet Eng Rev 25:41–76
3. Zhuang LL, Jin R, Zhu LH et al (2013) Promoter characterization and role of cAMP/PKA/CREB in the basal transcription of the mouse ORMDL3. Plos One 8(4):e60630. doi:10.1371/journal.pone.0060630
4. Keen JC, Cianferoni A, Florio G et al (2005) Characterization of a novel PMA-inducible pathway of interleukin-13 gene expression in T cells. Inmunology 117:29–37
5. Binia A, Van Stiphout N, Liang L et al (2013) A polymorphism affecting MYB binding within the promoter of the *PDCD4* gene is associated with severe asthma in children. Hum Mutat 34:1131–1139
6. Hu M, Ou-Yang HF, Wu CG et al (2014) Notch signaling regulates col1α1 and col1α2 expression in airway fibroblasts. Exp Biol Med 239:1589–1596
7. Ou-Yang HF, Wu CG, Qu S et al (2013) Notch signaling downregulates MUC5AC expression in airway epithelial cells through Hes1-dependent mechanisms. Respiration 86:341–346
8. Ni ZH, Tang JH, Cai ZY et al (2011) A new pathway of glucocorticoid action for asthma treatment through the regulation of PTEN expression. Respir Res 12:47
9. Matsukura S, Kurokawa M, Homma T et al (2013) Basic research on virus-induced asthma exacerbation: inhibition of inflammatory chemokine expression by fluticasone propionate. Int Arch Allergy Immunol 161(Suppl 2):84–92
10. Tanner FC, Carr DP, Nabel GJ et al (1997) Transfection of human endothelial cells. Cardiovasc Res 35:522–528
11. Castro FO, Portelles Y (1997) Transfección de ADN a células de mamíferos. Biotecnol Apl 14:149–161
12. Schenborn E, Groskreutz D (1999) Reporter gene vectors and assays. Mol Biotechnol 13:29–44
13. Köhler S, Belkin S, Schmid RD (2000) Reporter gene bioassays in environmental analysis. Fresenius J Anal Chem 366:769–779
14. Wet JR, Wood KV, Helinski DR et al (1985) Cloning of firefly luciferase cDNA and the expression of active luciferase in *Escherichia coli*. ProcNatl Acad Sci U S A 82:7870–7873
15. Dual-lucifersae reporter assay system promega. Technical Manual. Promega Corporation. http://www.promega.com
16. Hou Y, Cheng B, Zhou M et al (2014) Searching for synergistic bronchodilators and novel therapeutic regimens for chronic lung diseases from a traditional Chinese medicine, Qingfei Xiaoyan Wan. Plos One 9(11), e113104
17. Haque R, Hakim A, Moodley T et al (2013) Inhaled long-acting β2 agonists enhance glucocorticoid receptor nuclear translocation and efficacy in sputum macrophages in COPD. J Allergy Clin Immunol 132:1166–1173

18. Harrop CA, Gore RB, Evans CM et al (2013) TGF-β_2 decreases baseline and IL-13-stimulated mucin production by primary human bronchial epithelial cells. Exp Lung Res 39:39–47
19. Park IH, Um JY, Cho JS et al (2014) Histamine promotes the release of interleukin-6 via the H1R/p38 and NF-κB pathways in nasal fibroblasts. Allergy Asthma Immunol Res 6:567–572
20. Haj-Selem I, Fakhfakh R, Bérubé J et al (2014) MicroRNA-19a enhances proliferation of bronchial epithelial cells by targeting TGFβR2 gene in severe asthma. Allergy. doi:10.1111/all.12551
21. Garbacki N, Di Valentin E, Huynh-Thu VA et al (2010) MicroRNAs profiling in murine models of acute and chronic asthma: a relationship with mRNAs targets. Plos One 6(1):e16509
22. Takyar S, Vasavada H, Zhang J et al (2013) VEGF controls lung Th2 inflammation via the miR-1-Mpl (myeloproliferative leukemia virus oncogene)-P-selectin axis. J Exp Med 210:1993–2010
23. Jude JA, Dileepan M, Kannan MS et al (2012) miR-140-3p regulation of TNF-α-induced CD38 expression in human airway smooth muscle cells. Am J Physiol Lung Cell Mol Physiol 303:460–468
24. 260/280 and 260/230 Ratios. Technical bulletin. Thermo Fisher Scientific.http://www.nanodrop.com/Library/T009-NanoDrop%201000-&-NanoDrop%208000-Nucleic-Acid-Purity-Ratios.pdfAccessed. Accessed 17 Sept 2014
25. Colosimo A, Goncz KK, Holmes AR et al (2000) Transfer and expression of foreign genes in mammalian cells. Biotechniques 29:314–331
26. Franco AY, Longart M (2009) Aplicaciones de la proteína verde fluorescente (GFP) en la biología celular y en la visualización del sistema nervioso. Revista de Estudios Transdisciplinarios 1:84–96
27. Yang T, Sinai PS, Kitts PA et al (1997) Quantification of gene expression with a secreted alkaline phosphatase reporter system. Biotechniques 23:1110–1114
28. Ghim CM, Lee SK, Takayama S et al (2010) The art of reporter proteins in science: past, present and future applications. BMB Rep 43:451–460

Chapter 15

Review of Mouse Models Applied to the Study of Asthma

Fernando Marqués-García and Elena Marcos-Vadillo

Abstract

The diversity of asthma phenotypes increases its complexity. Animal models represent a useful tool to elucidate the pathophysiological mechanisms involved in both allergic and nonallergic asthma, as well as to identify potential targets for the development of new treatments. Among all available animal models, mice offer significant advantages for the study of asthma. In this chapter, the applications of mouse models to the study of asthma will be reviewed.

Key words Animal models, Asthma, Challenge, Mouse, Sensitization

1 Introduction

Asthma is a complex and heterogeneous disease characterized by chronic inflammation of the airway, hyperresponsiveness, and recurrent symptoms such as sneezing, coughing, and breathing difficulty [1]. This complexity is caused by multiple environmental as well as genetic factors [1], which defines different forms of the disease expression as phenotypes (observable clinics characteristics) and endotypes (different pathogenic mechanisms) [2].

Currently, multiple animal models have been developed, not only for the study of asthma but also for other allergic diseases such as atopic dermatitis [3], allergic conjunctivitis [4], food allergy, anaphylaxis [5], and allergic rhinitis [6]. These mouse models are important to elucidate the pathophysiological mechanisms of asthma, as well as to evaluate both safety and efficacy of therapies in preclinical phase, before starting clinical phases in humans. In this chapter, we will focus on the study of different mouse models applied to asthma.

María Isidoro-García (ed.), *Molecular Genetics of Asthma*, Methods in Molecular Biology, vol. 1434,
DOI 10.1007/978-1-4939-3652-6_15,

2 Mechanism of Asthma in Mouse

Animal models of asthma have been extensively used to examine mechanisms of the disease. Many advances in the understanding of the pathophysiology of asthma are based on these models [7]. The different cytokine profiles associated to asthma were initially described in mice [8]. Indeed, most of the mechanisms that are discussed today derive from studies conducted in animal models. As an example, the classical Th2 paradigm involving interleukin-4 (IL-4) or interleukin-5 (IL-5) was discovered using animal models [9]. However, these models only reflected allergic asthma. The identification of the interleukin 17 (IL-17), neutrophils that participate in the severe asthma or steroid resistant asthma, suggested that asthmatic disease was much more complex than what could be described by via Th1/Th2 paradigm [10]. These findings needed the development of appropriate mouse models that simulate the characteristics of these forms of asthma.

Respiratory viral infections can trigger asthma. The sendai virus (related to the human parainfluenza virus) has been administered to reach a chronic lung disease associated with airway hyperreactivity (AHR) in mouse [1]. Asthma is also triggered by pollution. Mouse models have been exposed to ozone (a major component of aerial pollution) to develop AHR [11].

For the study of intrinsic asthma, the strain A/J mice has been used. These mice spontaneously develop AHR without manipulation [12]. The intrinsic AHR has been associated to a chromosomal region containing *adam33* gene. This association was firstly identified in a mouse model [13]. The identification of genes associated with asthma in mice before they do in humans reveals that animal models may be useful for the study of human asthma.

In summary, these models allow us to study specific pathways or genes related to different forms of asthma. This reductionist approach greatly simplifies the study of such a heterogeneous disease, enabling the understanding of mechanisms that would otherwise be difficult to elucidate.

3 Is Mouse the Ideal Animal Model?

A wide variety of animal species have been used for the study of asthma [14]. Mouse, horse, rat, dog, sheep or monkey, have been used for studies of airway inflammation [14–17]. Each has advantages and disadvantages as asthma model (Table 1). Besides the mouse, the species most used for experimentation in asthma are the guinea pig [18], sheep [19], and monkey [20].

Despite the variety of available animal, mouse is the most used model. The widespread use of the mouse for studies of asthma is

Table 1
Advantages and disadvantages of different animal models for the study of asthma

	Advantages	Disadvantages
Mouse	Short gestational period Easy manipulation Reaction IgE-mediated Small and cheap	Not spontaneous AHR Anatomic differences Limited airway musculature
Rat	Reaction IgE-mediated Late airway response	Immunological reagents not abundant Need adjuvants for sensitization
Rabbit	Reaction IgE-mediated Late airway response	Difficult manipulation

due to the advantages over other animals. The choice of the mouse as a basic model is mainly based on scientific and economic reasons.

3.1 Scientific Reasons

Mouse is the most used model in multiple human diseases. This is due to the extensive knowledge we have obtained from the multiple genetic studies previously conducted [19] as well as the ease handling for the generation of transgenic animals [21]. With regard to allergic diseases, these animals are easily sensitized, using allergens such as the ovalbumin (OVA), and House Dust Mite (HDM) [22], or molds [23].

In addition, the existence of different mouse strains, which do not behave in the same way from the same allergen, is an advantage for identifying the mechanisms [24, 25] of inflammation and airway hyperresponsiveness.

In recent years, the rapid expansion of transgenic technology has allowed the development of mouse models in which the selective expression of a gene is inhibited (Knock-out), or induced (Knock-in) [26]. These tools allow us to more clearly understand the molecular pathways involved in the development of asthma [27].

3.2 Economics Reasons

Besides the previously exposed scientific reasons, there are also economic reasons that favor the use of mice as asthma model. Firstly, there are numerous and relatively cheap commercially available mouse-specific probes for studying allergic outcomes that allow conducting large studies. In addition, due to the small size and easily handling, many mice can be maintained in small areas.

Mouse also has a short life cycle. Its gestation period is 21 days that facilitates the rapid procurement of animals for experiments. In this short period, it has large litters (6–8 mice) that rapidly provide a lot of animals. In addition, it reaches sexual maturity in a relatively short period of time (6–8 weeks).

In summary, mouse appears to be a good model for conducting a variety of experiments aimed at elucidating the mechanisms involved in asthma.

4 Limitations of Animal Models in Asthma

Despite the undoubted advantages of animal models for the study of asthma, they also have limitations that must not be forgotten. These limitations must be always taken into account before choosing the model due to the influence that may have on the results obtained. The limitations relate to data extrapolation to humans, adjuvants used, chronicity of the disease, or anatomical differences, among others.

4.1 Extrapolation to Humans

The question that arises after obtaining data on animal models of diseases, including asthma, is how to extrapolate the results to humans. Animal models used in the laboratory do not spontaneously develop asthma; this is why different protocols are performed. The technique variations are particularly important as can have a significant impact on the outcome and are often overlooked when comparing responses between studies.

The general technique for inducing asthma in animals involves sensitizing to a previously unseen antigen and subsequently challenging the airways with the same antigen, in order to study the cascade of events or an outcome parameter of interest [28]. This is an experimental asthma induced in the airway of the animal that has to be compared to the naturally developed human asthma.

4.2 Adjuvants

Besides the allergen, an adjuvant is usually added [8]. This molecule modulates the immune response. It acts as immune enhancer, ensuring a sufficiently intense immune response. One of the more used adjuvants is aluminum hydroxide or Alum [29]. Other adjuvants, used although to a lesser proportion, are heat-killed *Bordetella pertussis* [30] or the complete Freund's adjuvant [28].

The main problem of using adjuvants is that they may alter the mechanism of sensitization to the allergen under consideration, and modify the immune response [26]. To avoid these problems, adjuvant-free models [31] or models that inoculate previously stimulated immune cells such as T lymphocytes [32] have been developed.

4.3 Chronicity of Asthma

Besides the inflammatory process, a remodeling of the airway occurs as a result of the chronicity in asthma. This remodeling involves goblet cell metaplasia and hyperplasia [33], mucus hyper secretion [34], and thickening of airway smooth muscle [35] due to the repeated exposure to the allergen. In addition, animal models initially respond to the intranasal allergen provocation, but when provocation is prolonged, the animal may develop tolerance [36]. To avoid this, the provocation can be done with low doses maintained over time. With this strategy, mouse models that express the characteristics of chronic asthma have been developed [37].

4.4 Asthma in Early Life

The animals used for the study of asthma are normally adults. However, human asthma can appear early in life. At that point, there are situations such as in uterus environment [38], viral infection [39], exposure to allergens [40], smoking, and pollution [41] or pets [42] that should be considered, therefore, models that allow studying asthma in early life are needed.

4.5 Anatomical Differences

It is important to highlight that the anatomical structure of the airway is different in rodents than humans. Animals are quadrupeds and due to gravity, this position may influence the effort made by the lungs to inside moving the air [43]. This situation is aggravated by the airflow limitation due to the asthma reaction.

Another important aspect is the morphology and arrangement of the bronchial tree that affects the penetration of the allergen into the lung. It has been reported an inverse relationship between body size and relative airway caliber in rodents [44]. In addition, mice exhibit a thin smooth muscle layer that causes an easier constriction of the airway [45] and a high number of goblet cells [46].

4.6 Size

The size of the animal greatly influences lung function. For long time, to evaluate lung function in small animals has been challenging. Currently there are three methods:

- Noninvasive methods: the most widely used is whole body plethysmography. The animal is placed in a chamber and the respiratory parameters are indirectly analyzed before and after methacholine administration.
- Invasive methods: it is considered the gold standard. The animal is anesthetized and a tracheal tube is introduced to directly measure lung volumes before and after the methacholine administration.
- Electrical field stimulation (EFS): although the in vivo response to inhaled methacholine is the most widely used method of assessing the AHR, it was initially limited due to the difficulty of delivering an aerosol to the mouse airways. EFS has been described as an alternative method to assess AHR in mouse models.

5 Asthma Model Design

Currently, there are a variety of mouse models for the study of asthma. The general outline is based on an initial systemic sensitization followed by an aerial local elicitation. Depending on the hypothesis, several aspects must be taken into account, the mouse strain, type of allergen, route of administration, and induction time among others.

5.1 Mouse Strain

The availability of a high number of different mouse strains is an advantage. Mouse strains can be classified according to the capacity of developing airway inflammation and AHR. There are responder strains, such as A/J and AKR/J with high levels of AHR to methacholine [47] or non-responder strains, as the C3H/HeJ or DBA/2, that are resistant to allergen-induced AHR [48]. However, the most used are the BALB/c and C57B/6 strains because their immune response is well characterized. The BALB/c respond occurs via Th2 that typically induces allergic parameters, such as IgE production, AHR, and eosinophilic inflammation of the airway; however, C57CL/6 has limitations in developing allergic airway response because the immune response occurs via Th1. This strain is used in many genetically manipulated mice.

Different behaviors of mouse strains are mainly due to genetic characteristics. Genetic manipulation has favored the generation of mice to study molecular mechanisms involved in asthma. Thus, mice that do not express a particular gene (knock out) or conversely overexpress the gene of interest (knock in) have been developed. A third model is the conditional knockout mouse in which the expression of the gene can be manipulated to the necessary extent.

5.2 Type of Allergen

A variety of allergens have been used in animal models. One of the most used allergen has been ovalbumin (OVA) in both sensitization and challenge phases. It is cheap, well characterized, and can be produced in large quantities [49]. As already described, the continuous administration of the allergen can trigger tolerance in various mouse strains. At the stage of sensitization, OVA is usually injected with an intraperitoneal adjuvant. The challenge phase is performed by air without adjuvant. The allergic induction caused by OVA is not the same to be obtained by allergens usually present in the environment.

The generation of asthma models that are more approximate to human requires allergens such as pollen, mods, or HDM, naturally present in the environment. These aeroallergens, suspended in the air, directly reach the airway, while OVA normally access to the body through the digestive tract. The progress of biotechnological techniques is focused in generating isolated epitopes responsible for the asthmatic phenotype to achieve more specific and potent responses than the obtained with extracts. Other models combine two or more allergens getting stronger inflammatory responses [50]. Finally, allergen concentration varies depending on the phase of the reaction and type of allergen.

5.3 Route of Allergen Administration

The route of allergen administration depends on the phase of the experiment. Thus, in the sensitization phase, the intraperitoneal via is commonly used, while in the provocation phase the allergen is introduced by air to generate a local response. Nowadays, in the sensitization phase, it is common to replace the intraperitoneal route by the intubation, resembling what happens in human asthma [51].

There are different strategies in the elicitation phase. Nebulization implies that the concentration of allergen does not penetrate far enough into the airway remaining in the upper respiratory tract. Other possibility is the application of allergen directly in the airway [52, 53], on the nostrils of the mouse, via intra-tracheal or depositing the allergen into the lungs with a bronchoscope. The choice of the method to be used is determined by the availability of material, as well as technical skill at the laboratory.

5.4 Induction Time

In mice, the induction of asthma can be achieved in short exposures to the allergen, days or weeks, or in longer periods of months. According to the exposure time, there are two different models, acute or short-term and chronic or long-term model. In acute models, high concentrations of allergen are used to obtain the asthmatic response in a short period. These models are useful to study airway inflammation but do not reflect all changes that occur in the human response.

However, in chronic models, the exposure to the allergen for longer periods produces inflammation and airway remodeling as in human asthma. The main problem of chronic models is tolerance that can be avoided by using low doses of allergen, as previously mentioned.

6 Mouse Models of Asthma

Most mouse strains do not spontaneously develop allergic airway inflammation or AHR; for this reason, different inducing agents are used. The mouse models of asthma can be grouped according to the phenotypes of human asthma. The models that have been conventionally developed are those aimed to study allergic asthma in acute or chronic models as previously seen.

Other models to study asthma phenotypes have been designed. In these models, molecules like ozone, cigarette smoke, diesel particles, or infectious agents can induce the response as respiratory virus although models of intrinsic asthma such as the strain A/J can be used. In addition to these classic models, genetically modified mice help to better understand the metabolic pathways involved in asthma.

Finally, different types of purified blood cells from asthmatic patients can be transferred to nude mice SCID (Severe Combined Immunodeficiency) simulating human asthmatic reactions in mice. All these models can also be used for the identification of new therapeutic targets for the disease as well as for the development of new drug.

In summary, the mouse models represent an opportunity to study the mechanisms involved in asthma and find targets to develop new treatments. The challenge in this field is to develop models that more closely approximate to human asthma, reflecting all changes that occur in the disease.

Acknowledgments

This work was supported by grants of the Junta de Castilla y León ref. GRS1047/A/14 and GRS1189/A/15.

References

1. Kim HY, DeKruyff RH, Umetsu DT (2010) The many paths to asthma: phenotype shaped by innate and adaptive immunity. Nat Immunol 11(7):577–584
2. Agache I, Akdis C, Jutel M et al (2012) Untangling asthma phenotypes and endotypes. Allergy 67:835–846
3. Jin H, He R, Oyoshi M et al (2009) Animal models of atopic dermatitis. J Invest Dermatol 129:31–40
4. Niederkorn JY (2008) Immune regulatory mechanisms in allergic conjunctivitis: insights from mouse models. Curr Opin Allergy Clin Immunol 8:472–476
5. Dearman RJ, Kimber I (2007) A mouse model for food allergy using intraperitoneal sensitization. Methods 41:91–98
6. Wagner JG, Harkema JR (2007) Rodent models of allergic rhinitis: relevance to human pathophysiology. Curr Allergy Asthma Rep 7:134–140
7. Shin YS, Takeda K, Gelfand EW et al (2009) Understanding asthma using animal models. Allergy Asthma Immunol Res 1:10–18
8. Kips JC, Anderson GP, Fredberg JJ et al (2003) Murine models of asthma. Eur Respir J 22: 374–382
9. Holt PG, Macaubas C, Stumbles PA et al (1999) The role of allergy in the development of asthma. Nature 402:B12–B17
10. Maddox L, Schwartz DA (2002) The pathophysiology of asthma. Annu Rev Med 53: 477–498
11. Robays LJ, Maes T, Joos GF et al (2009) Between a cough and a wheeze: dendritic cells at the nexus of tobacco smoke-induced allergic airway sensitization. Mucosal Immunol 2:206–219
12. Li N, Hao M, Phalen RF et al (2003) Particulate air pollutants and asthma. A paradigm for the role of oxidative stress in PM-induced adverse health effects. Clin Immunol 109:250–265
13. Johnston RA, Zhu M, Rivera-Sanchez YM et al (2007) Allergic airway responses in obese mice. Am J Respir Crit Care Med 176:650–658
14. Kim EY, Battaile JT, Patel AC et al (2008) Persistent activation of an innate immune response translates respiratory viral infection into chronic lung disease. Nat Med 14:633–640
15. Pichavant M et al (2008) Ozone exposure in a mouse model induces airway hyperreactivity that requires the presence of natural killer T cells and IL-17. J Exp Med 205:385–393
16. Hadeiba H, Corry DB, Locksley RM (2000) Baseline airway hyperreactivity in A/J mice is not mediated by cells of the adaptive immune system. J Immunol 164:4933–4940
17. Van Eerdewegh P, Little RD, Dupuis J et al (2002) Association of the ADAM33 gene with asthma and bronchial hyperresponsiveness. Nature 418:426–430
18. Haitchi HM, Bassett D, Bucchieri F et al (2009) Induction of a disintegrin and metalloprotease 33 during embryonic lung development and the influence of IL-13 or maternal allergy. J Allergy Clin Immunol 124:590–597
19. Fairbairn SM, Page CP, Lees P et al (1993) Early neutrophil but not eosinophil or platelet recruitment to the lungs of allergic horses following antigen exposure. Clin Exp Allergy 23:821–828
20. Colasurdo GN, Hemming VG, Prince GA et al (1998) Human respiratory syncytial virus produces prolonged alterations of neural control in airways of developing ferrets. Am J Respir Crit Care Med 157:1506–1511
21. Toward TJ, Broadley KJ (2004) Early and late bronchoconstriction, airway hyperreactivity, leukocyte influx and lung histamine and nitric oxide after inhaled antigen: effects of dexa-

methasone and rolipram. Clin Exp Allergy 34:91–102
22. Chen W, Alley MR, Manktelow BW (1991) Airway inflammation in sheep with acute airway hypersensitivity to inhaled Ascaris suum. Int Arch Allergy Appl Immunol 96:218–223
23. Johnson HG, Stout BK (1993) Late phase bronchoconstriction and eosinophilia as well as methacholine hyperresponsiveness in Ascaris-sensitive rhesus monkeys were reversed by oral administration of U-83836E. Int Arch Allergy Immunol 100:362–366
24. Dietrich WF, Miller J, Steen R et al (1996) A comprehensive genetic map of the mouse genome. Nature 380:149–152
25. Elias JA, Lee CG, Zheng T et al (2003) New insights into the pathogenesis of asthma. J Clin Invest 111:291–297
26. Fattouh R, Pouladi MA, Alvarez D et al (2005) House dust mite facilitates ovalbumin-specific allergic sensitization and airway inflammation. Am J Respir Crit Care Med 172:314–321
27. Mehlhop PD, VandeRijn M, Goldberg AB et al (1997) Allergen induced bronchial hyperreactivity and eosinophilic inflammation occur in the absence of IgE in a mouse model of asthma. Proc Natl Acad Sci U S A 94:1344–1349
28. Zosky GR, Sly PD (2007) Animal models of asthma. Clin Exp Allergy 37:973–988
29. Whitehead GS, Walker JKL, Berman KG et al (2003) Allergen-induced airway disease is mouse strain dependent. Am J Physiol 285:L32–L42
30. WillsKarp M, Ewart SL (1997) The genetics of allergen-induced airway hyperresponsiveness in mice. Am J Respir Crit Care Med 156:S89–S96
31. Brewer JM, Conacher M, Hunter CA et al (1999) Aluminium hydroxide adjuvant initiates strong antigen-specific Th2 responses in the absence of IL-4 or IL-13 mediated signaling. J Immunol 163:6448–6454
32. Schneider T, van Velzen D, Moqbel R et al (1997) Kinetics and quantitation of eosinophil and neutrophil recruitment to allergic lung inflammation in a brown Norway rat model. Am J Respir Cell Mol Biol 17:702–712
33. Nakagome K, Dohi M, Okunishi K et al (2005) Antigen-sensitized CD4(1)CD62l(low) memory/effector T helper 2 cells can induce airway hyperresponsiveness in an antigen free setting. Respir Res 6:46
34. Farraj AK, Harkema JR, Jan TR et al (2003) Immune responses in the lung and local lymph node of A/J mice to intranasal sensitization and challenge with adjuvant-free ovalbumin. Toxicol Pathol 31:432–447
35. Hogan SP, Koskinen A, Matthaei Ki et al (1998) Interleukin-5 producing CD4(1) T cells play a pivotal role in aeroallergen-induced eosinophilia, bronchial hyperreactivity, and lung damage in mice. Am J Respir Crit Care Med 157:210–218
36. Rose MC, Voynow JA (2006) Respiratory tract mucin genes and mucin glycoproteins in health and disease. Physiol Rev 86:245–278
37. Young HWJ, Sun CX, Evans CM et al (2006) A adenosine receptor signaling contributes to airway mucin secretion after allergen challenge. Am J Respir Cell Mol Biol 35:549–558
38. James A (2005) Remodelling of airway smooth muscle in asthma: what sort do you have? Clin Exp Allergy 35:703–707
39. Kumar RK, Foster PS (2002) Modeling allergic asthma in mice—pitfalls and opportunities. Am J Respir Cell Mol Biol 27:267–272
40. Temelkovski J, Hogan SP, Shepherd DP et al (1998) An improved murine model of asthma: selective airway inflammation, epithelial lesions and increased methacholine responsiveness following chronic exposure to aerosolised allergen. Thorax 53:849–856
41. Gern JE, Lemanske RF, Busse WW (1999) Early life origins of asthma. J Clin Invest 104:837–843
42. Holt PG, Sly PD (2002) Interactions between RSV infection, asthma, and atopy: unraveling the complexities. J Exp Med 196:1271–1275
43. Illi S, von Mutius E, Lau S et al (2006) Perennial allergen sensitisation early in life and chronic asthma in children: a birth cohort study. Lancet 368:763–770
44. Persson CGA, Erjefalt JS, Korsgren M et al (1997) The mouse trap. Trends Pharmacol Sci 18:465–467
45. Celedon JC, Litonjua AA, Ryan L et al (2002) Exposure to cat allergen, maternal history of asthma, and wheezing in first 5 years of life. Lancet 360:781–782
46. Bettinelli D, Kays C, Bailliart O et al (2002) Effect of gravity on chest wall mechanics. J Appl Physiol 92:709–716
47. Gomes RFM, Shen X, Ramchandani R et al (2000) Comparative respiratory system mechanics in rodents. J Appl Physiol 89:908–916
48. Karol MH (1994) Animal models of occupational asthma. Eur Respir J 7:555–568
49. Ewart SL, Kuperman D, Schadt E et al (2000) Quantitative trait loci controlling allergen-induced airway hyperresponsiveness in inbred mice. Am J Respir Cell Mol Biol 23:537–545
50. McIntire JJ, Umetsu SE, Akbari O et al (2001) Identification of Tapr (an airway hyperreactivity regulatory locus) and the linked Tim gene family. Nat Immunol 2:1109–1116

51. Fuchs B, Braun A (2008) Improved mouse models of allergy and allergic asthma—chances beyond ovalbumin. Curr Drug Targets 9:495–502
52. Sarpong SB, Zhang LY, Kleeberger SR (2003) A novel mouse model of asthma. Int Arch Allergy Immunol 32:346–354
53. Cates EC, Gajewska BU, Goncharova S et al (2003) Effect of GM-CSF on immune, inflammatory, and clinical responses to ragweed in a novel mouse model of mucosal sensitization. J Allergy Clin Immunol 111: 1076–1086

Chapter 16

General and Specific Mouse Models for Asthma Research

Fernando Marqués-García and Elena Marcos-Vadillo

Abstract

To study the complexity of human asthma disease, the development of different animal models is needed. Among all different laboratory animals, mice represent a useful tool for the development of asthma. This chapter will describe protocols for designing different animal models applied to the studying of asthma phenotypes.

Key words Acute, Allergen, Chronic, Mouse, Intrinsic, Occupational

1 Introduction

Human asthma is a heterogeneous and complex disease, in which both inflammation and remodeling of the airway structure take place [1]. The molecular and cellular mechanisms that trigger the disease are not entirely understood; therefore, animal models represent an alternative to investigate the mechanisms and progression of the disease. Due to the complexity of the disease, a single model is not sufficient to reproduce all changes [2]. Mouse species is widely used for generating such models. Different models have been generated in which the exposure can be acute or chronic, *see* Tables 1 and 2 for examples induced by ovalbumin (OVA).

The acute models are useful for studying the inflammatory response, whereas chronic models better replicate the clinical features of asthma as well as remodeling of the airway and bronchial hyperresponsiveness [9–11]. OVA is rarely involved in human allergic asthma processes, there are multiple models developed based on allergens causing human disease such as House Dust Mite (HDM) [12] pollen [13], fungal spores [14] etc. Also it has been worked in the generation of mouse models to study occupational asthma [15, 16] and intrinsic asthma [17]. In this chapter, different models for the study of asthma will be described.

María Isidoro-García (ed.), *Molecular Genetics of Asthma*, Methods in Molecular Biology, vol. 1434,
DOI 10.1007/978-1-4939-3652-6_16, © Springer Science+Business Media New York 2016

Table 1
Different treatment strategies in acute OVA mouse models for strain BALB/c

Sensitisation	Challenge	References
OVA/Alum (IP) 0 & 12 days	OVA aerosol 18–23 days	[3]
OVA/Alum (IP) 0 & 14 days	OVA aerosol 28–30 days	[4]
OVA/Alum (IP) 0 & 14 days	OVA IN 14,25,26 & 27 days	[5]

OVA ovalbumin, *Alum* aluminium hydroxide, *IP* intraperitoneal, *IN* intranasal

Table 2
Different treatment strategies in chronic OVA mouse models

Sensitisation	Challenge	References
OVA/Alum (IP) -7 & -21 days	OVA 6/8 weeks, 3 days/week	[6]
OVA/Alum (IP) 0 & 14 days	OVA IN 14, 27, 28, 47, 61 & 73 days	[7]
OVA/Alum (IP) 0 & 14 days	OVA IN 12 weeks, 3 days/week	[8]

Description of the differences in both stages in female strain BALB/c
OVA ovalbumin, *alum* aluminium hydroxide, *IP* intraperitoneal, *IN* intranasal

2 Materials

2.1 Animals

1. 6–8 Weeks old female BALB/c mice (Charles Rivers Lab).
2. 6–8 Weeks old female A/J old (Charles Rivers Lab).

2.2 Common Material

1. 1× PBS (Phosphate-Buffered saline): 1.37 mM NaCl, 2.6 mM KCl, 10 mM Na_2HPO_4, 1.8 mM KH_2PO_4. Adjust the pH to 7.4 with HCl.
2. Tuberculin syringes (BD Biosciences).
3. Needle 20G short (BD Biosciences).
4. Ketamine chlorhydrate (Imalgen, Merial).
5. Medetomidine hydrochloride (Medetor, Virbac).
6. 100 μl micropipette and appropriate tips (Thermo-Fischer).

2.3 OVA Asthma Model

1. Ovalbumin (Sigma-Aldrich).
2. Aluminum hydroxide (Sigma-Aldrich).

2.4 Allergen-Specific Asthma Models

2.4.1 Pollen

1. Crude olive pollen (*Olea europaea*) extract (ALK-Abello Laboratories).
2. Aluminum hydroxide (Sigma-Aldrich).

2.4.2 Dust/Mites

1. House Dust Mite (HDM) Extract (Allergopharma).
2. Freund´s adjuvant: 1 ml contains 1 mg of *Mycobacterium tuberculosis* (H37Ra, ATCC 25177), heat killed and dried, 0.85 ml paraffin oil and 0.15 ml mannide monooleate (Sigma-Aldrich).

2.4.3 Molds

1. *Alternaria alternata* (strain 18586).
2. Potato dextrose agar plates: Per liter, 200 g of infusion from potatoes, 20 g dextrose, and 15 g agar pH 5.6 (Difco).
3. Cell scraper.
4. Hemocytometer.
5. Czapek's medium: Per litre, 30 g sucrose, 2 g $NaNO_3$, 1 g K_2HPO_4, 0.5 g $MgSO_4$, 0.5 KCl, 0.01 $FeSO_4$, 15 g agar pH 7.3 (Himedia Labs).
6. NH_4HCO_3 + polyvinyl polypyrrolidone (Sigma-Aldrich).
7. Glycerol.
8. Aluminum hydroxide (Sigma-Aldrich).

2.5 Tobacco-Induced Asthma Models

1. Plexiglas chamber.
2. Nebulizer (Mega Medical).
3. Tobacco cigarettes.
4. *E. coli* LPS (Sigma-Aldrich).

3 Methods

3.1 Ovalbumin Asthma Model

Models based on the use of OVA are grouped into two types according to the duration of the treatment, as described below:

- Acute Model.
- Chronic Model.

3.1.1 Acute Model

The experiments are performed with 6–8 weeks old female BALB/c mice (*see* **Notes 1** and **2**) (Fig. 1). Animals must be kept confined and handled to meet the current ethics rules of handling animals for experimentation (*see* **Notes 3** and **4**). Start from a population of, for example, eight mice. The animals are divided into two groups:

- Group 1: Four mice that are not treated with OVA Intraperitoneal/Intranasal (in place 1× PBS is given).
- Group 2: Four mice are treated with OVA Intraperitoneal/Intranasal.

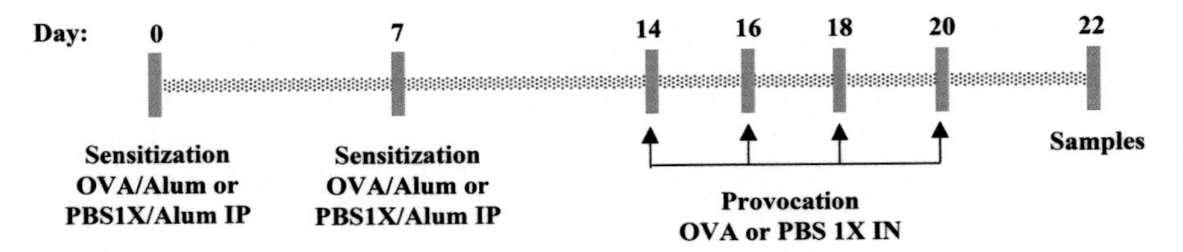

Fig. 1 Acute mouse model. *IP* intraperitoneal, *IN* intranasal, *PBS* phosphate-buffered saline, *OVA* ovalbumin, *Alum* aluminum hydroxide

1. On day 0 (start of experiment), inject the animals intraperitoneally with 20 μg of OVA emulsified in 0.2 ml 1× PBS containing 2 mg of aluminum hydroxide (these doses are administered to each mouse). Inject the controls with 0.2 ml of 1× PBS containing 2 mg of aluminum hydroxide (*see* **Note 5**).
2. Keep the animals stabled for 1 week (7 days) for the hypersensitization reaction to occur.
3. On day 7, administer the second dose of OVA to reinforce hyper-sensitization by repeating **step 1**.
4. Keep the mice housed for another 7 days.
5. After 14 days of hyper-sensitization, organize the provocative steps, four provocations in total.
6. Perform the first intranasal provocation with OVA.
7. Anesthetize the animal with mild sedation. Administer an intraperitoneal dose of Ketamine (75 mg/kg) + Medetomidine (1 mg/kg). Wait for the complete sedation of the animal before handling (*see* **Notes 6–9**).
8. Stand upright the sedated animal with his nose up and give 50 μl of OVA (0.1 % in 1× PBS) for the experimental group or 50 μl of 1× PBS for the control group. Administrate the dose dropping drop wise to the snout (helping with a micropipette), favoring inoculation during the inspiratory process of the animal.
9. Keep the animal in that position for 1–2 min.
10. Return the animal to the cage, put horizontally on its side for favoring ventilation (*see* **Note 10**).
11. Repeat the procedure described in **steps 7–10**, three times with a frequency of 2 days (48 h) between one provocation and the next. This protocol would correspond to day 16 (second provocation), 18 (third provocation), and 20 (fourth provocation) (*see* **Note 11**).
12. On day 22 of the experiment, proceed to the collection of the samples.
13. 2 Days (48 h) after the last provocation, perform the necropsy of the animals.

14. Anesthetize the animal with the same mixture described in paragraph 7 (*see* **Note 12**).
15. With the anesthetized animal, proceed to the necropsy. The mouse is placed in ***decubitus supino*** position fixing the legs to a surface.
16. For obtaining samples:
    - Blood is obtained by cardiac puncture at the ventricular level. This puncture serves as gathering blood process and as euthanasia method. Cellular and soluble factors present in serum can be analyzed.
    - BALF (Bronchoalveolar Lavage): Cannulate the trachea and inject 1 ml of 1× PBS, two times. BALF is used for studying cellular and soluble factors.
    - Tissues: Collect the organs of interest. They can be used for molecular biology as well as histological studies.
    - Hematopoietic organs: Can be employed for in vitro cell culture.

3.1.2 Chronic Model

The experiments are performed with 6–8 weeks old female BALB/c mice (*see* **Notes 1** and **2**) (Fig. 2). Animals must be kept confined and handled to meet the current ethics rules of handling animals for experimentation (*see* **Notes 3** and **4**). Start for example with 8 mice. The animals are divided into two groups:

- Group 1: Four mice that are not treated with OVA Intraperitoneal/Intranasal (in place 1× PBS is given).
- Group 2: Four mice are treated with OVA Intraperitoneal/Intranasal.

1. On day 0 (start of experiment), inject the animal intraperitoneally with 20 μg of OVA emulsified in 0.2 ml 1× PBS containing 2 mg of aluminum hydroxide (these doses are administered to each mouse). Inject the controls with 0.2 ml of 1× PBS containing 2 mg of aluminum hydroxide (*see* **Note 5**).
2. Keep the animal animals stabled for 1 week (7 days) for the OVA hyper-sensitization to occur.

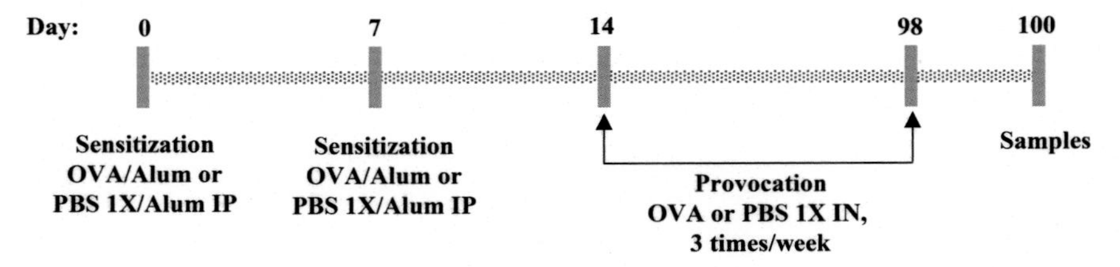

Fig. 2 Chronic mouse model. *IP* intraperitoneal, *IN* intranasal, *PBS* phosphate-buffered saline, *OVA* ovalbumin, *Alum* aluminum hydroxide

3. On day 7, administer the second dose to reinforce hyper-sensitization by repeating the **step 1**.
4. Keep the mice housed for another 7 days.
5. After 14 days of hyper-sensitization, organize the steps of provocation. In this model, the frequency is three provocations per week.
6. Perform the first intranasal provocation with OVA on day 14 of the experiment. Anesthetize the animal by administering an intraperitoneal dose of Ketamine (75 mg/kg) + Medetomidine (1 mg/kg). Wait for the complete sedation of the animal before you start handling (*see* **Notes 6–9**).
7. Stand upright the sedated animal with his nose up and give 50 μl of 1× PBS to the control group or 50 μl of OVA (0.1% in 1× PBS) to the experimental group. Administrate dropping drop wise to the snout (helping with a micropipette), favoring inoculation during the inspiratory process animal.
8. Keep the animal in that position for 1–2 min.
9. Return the animal to the cage, putting horizontally on its side for favoring ventilation (*see* **Note 10**).
10. In this chronic model, monitor the clinical course of the mice. Use a classification or Clinical Score that assigns a numerical value to the clinical signs developed for mouse (*see* **Note 13**).
11. Repeat the procedure described in **steps 6–9**, three times per week for 12 weeks (3 months).
12. Observe and record the signs observed in mice during this period. After that time, proceed to the collection of samples.
13. Proceed to the necropsy of the animals 2 days (48 h) after the last provocation.
14. Anesthetize the animal with the same mixture described in paragraph 6 (*see* **Note 12**).
15. Place it in ***decubitus supino*** fixing it to a surface by the legs.
16. For obtaining the samples:
    - Blood is obtained by cardiac puncture at the ventricular level. This puncture serves as gathering blood and euthanasia method. Cellular as well as soluble factors present in serum can be analyzed.
    - BALF (Broncho alveolar Lavage): Cannulate the trachea and inject 1 ml of 1× PBS 2 times. BALF is used for cellular and soluble factors studies.
    - Tissues: Collect the organs of interest for molecular biology as well as histological studies.
    - Collect the hematopoietic organs for in vitro cell culture.

3.2 Specific Allergen Asthma Model

In this section, we focus on describing asthma mouse models, where relevant allergens in humans are used to trigger allergic asthma. Models for the following allergens are described:

- Pollen.
- Dust/Mite.
- Molds.

3.2.1 Pollen

Use *Olea europaea* (Olive) as pollen allergen that is responsible for a high percentage of seasonal allergic asthma in the Mediterranean region (Fig. 3). The experiments are performed with 6–8 weeks old female BALB/c mice.

The mice are divided into two groups:

- Control: Inoculated with 1× PBS.
- Experimental: Inoculated with pollen allergen.

1. Prepare the mixture of 25 μg pollen extract with 50 mg of aluminum hydroxide in 200 μl of 1× PBS for each mouse and inoculation.
2. On day 0, inoculate subcutaneously the control group with 200 μl of 1× PBS and the experimental group with 200 μl of the mixture prepared in **step 1**.
3. During the stage of hyper-sensitization, inoculate once a week for the first 3 weeks.
4. From week 3 to week 10 do not treat the animals.
5. At week 10, perform the elicitation phase after 72 h (3 days) and 48 h (2 days) before slaughter. Perform the first challenge 3 days before slaughter of animals.
6. Inject an intraperitoneal dose of ketamine (75 mg/kg) + Medetomidine (1 mg/kg) mixture to the animal.
7. Place the sedated animal horizontally with the muzzle up and inoculate (using a micropipette) 30 μl of pollen extract (15 μg of extract of pollen per mouse and provocation) to the experimental group and 30 μl of 1× PBS to the control group.
8. Repeat **steps 6** and 7 for the second provocation, 2 days before the necropsy of the mice.

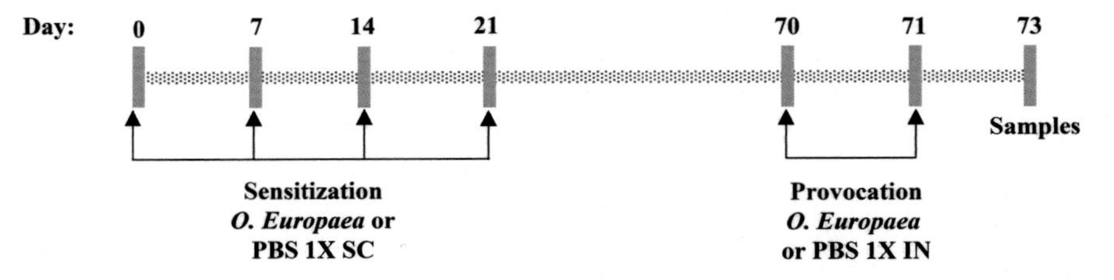

Fig. 3 Specific allergen asthma model: Polen. *SC* sub-cutaneus, *IN* Intranasal, *PBS* phosphate-buffered saline

9. At 48 h, proceed to the necropsy of the animals for the collection of samples.
10. Anesthetize the animal with the same mixture described in **step 6**.
11. With the anesthetized animal, proceed to the necropsy, setting the animal in decubitus supino position.
12. Obtain the samples (blood, BALF, tissues, and/or hematopoietic organs).

3.2.2 House Dust Mite (HDM)-Induced Asthma Model

Protocol 1

Use extract of house dust mite (HDM) as allergen. It is responsible for a high percentage of non-seasonal allergic asthma worldwide (Fig. 4). The experiments are performed with 6–8 weeks old female BALB/c mice.

1. Dissolve the HDM in 1× PBS at a concentration of 100 μg of extract in 50 μl of 1× PBS (without adjuvant) per mouse of the experimental group. Give 50 μl of 1× PBS to the controls.
2. Start the experiment (day 0), perform the first administration of HDM extract or PBS, depending on the group. Give the animals an intraperitoneal dose of ketamine (75 mg/kg) + Medetomidine (1 mg/kg) mixture. Perform the allergen deposition intranasal.
3. Repeat the days 7, 14, and 21 of the experiment.
4. On day 23 (48 h after the last allergen exposure) proceed to the collection of samples.
5. Anesthetize the animal with the same mixture described in **step 2**.
6. With the anesthetized animal, perform the necropsy, setting the animal in *decubitus supino* position to obtain samples (blood, BALF, tissues, and/or hematopoietic organs).

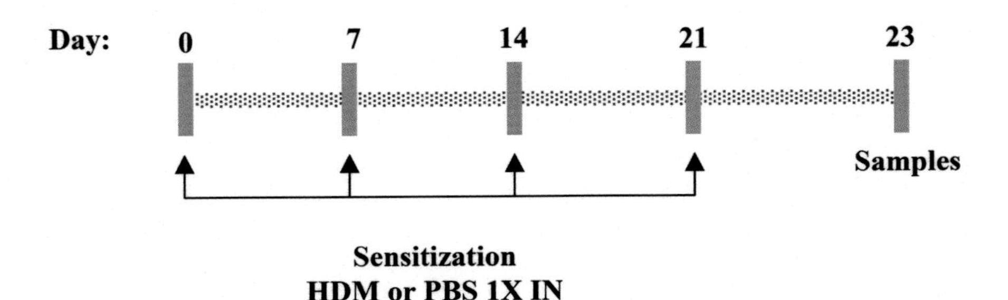

Fig. 4 Specific allergen asthma model: house dust mite (HDM)-induced asthma model protocol 1. *IN* intranasal, *PBS* phosphate-buffered saline, *HDM* house dust mite

Protocol 2

1. In this protocol, the allergen extract of house dust mite (HDM) at a concentration ten times lower than that used in protocol 1 (Fig. 5) and an adjuvant (aluminum hydroxide) are used.
2. Dissolve the HDM in 1× PBS at a concentration of 10 μg of extract plus 0.5 mg aluminum hydroxide in 50 μl of 1× PBS per mouse in the experimental group. Give 50 μl of 1× PBS with 0.5 mg of aluminum hydroxide to the control mice.
3. Prepare the 6–8 weeks old female BALB/c mice.
4. Start the experiment (day 0) with the first administration of HDM extract or 1× PBS (plus adjuvant for both), depending on the group. In this case, it is administrated by intraperitoneal pathway.
5. Maintain the mice housed for 14 days.
6. On day 14, repeat the **step 4**.
7. Maintain the mice housed for 7 days more.
8. The HDM is prepared at a concentration of 10 μg in 50 μl of extract in 1× PBS without aluminum hydroxide.
9. On day 21, perform the first provocation by administering the allergen prepared in section 8, intranasal.
10. Inject an intraperitoneal dose of ketamine (75 mg/kg) + Medetomidine (1 mg/kg) mixture.
11. The allergen deposition is intranasal.
12. Repeat **steps 10** and **11** on days 22 and 23 of the experiment.
13. On day 25 (48 h after the last allergen exposure), proceed to the collection as samples.
14. Anesthetize the animal with the same mixture described in **step 10**.
15. With the anesthetized animal, proceed to the necropsy setting the animal in ***decubitus supino*** position for obtaining samples (blood, BALF, tissues, and/or hematopoietic organs).

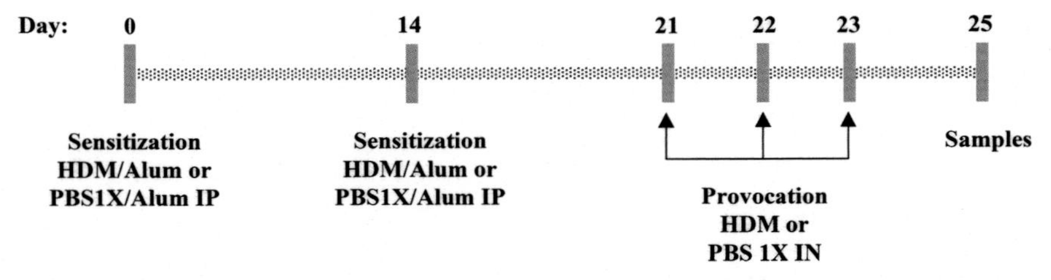

Fig. 5 Specific allergen asthma model: house dust mite (HDM)-induced asthma model protocol 2. *IP* intraperitoneal, *IN* intranasal, *PBS* phosphate-buffered saline, *HDM* house dust mite, *Alum* aluminum hydroxide

3.2.3 House Dust Mite (HDM)-Induced Severe Asthma Model

1. In this protocol, the complete Freund's adjuvant (Fig. 6) is used along with an allergen extract of house dust mite (HDM).
2. The HDM is dissolved in 1× PBS at a concentration of 100 μg of more adjuvant extract in 50 μl of PBS 1× per mouse in the experimental group. Give 50 μl of 1× PBS plus adjuvant to the control mice.
3. The experiments are performed with 6–8 weeks old female BALB/c mice.
4. Perform the first administration of HDM extract or PBS 1× (plus adjuvant for both), depending on the group (sensitization phase) on day 0. In this case, administer by sub-cutaneous pathway.
5. Maintain the mice housed for 14 days.
6. Prepare the HDM at a concentration of 100 μg of extract in 50 μl of 1× PBS.
7. On day 14, perform a single provocation by administering the allergen prepared by via intranasal.
8. Inject an intraperitoneal dose of ketamine (75 mg/kg) + Medetomidine (1 mg/kg) mixture.
9. Perform the allergen deposition intranasal.
10. On day 15 (24 h after the last allergen exposure), proceed to the collection of samples.
11. The animal is anesthetized with the same mixture described in **step 8**.
12. With the anesthetized animal, proceed to the necropsy setting the animal in *decubitus supino* position for obtaining samples (blood, BALF, tissues, and/or hematopoietic organs).

3.2.4 Molds

1. Use extract of spores of the fungus *Alternaria alternata* (strain 18586) as allergen.
2. These molds are cultured at 27 °C on potato dextrose agar plates for 1 week, before gently harvesting the spores with a cell scraper.
3. Dilute the spores in 1× PBS for counting with a hemocytometer and resuspend in 1× PBS at a concentration of $2–3 \times 10^7$ spores/ml.

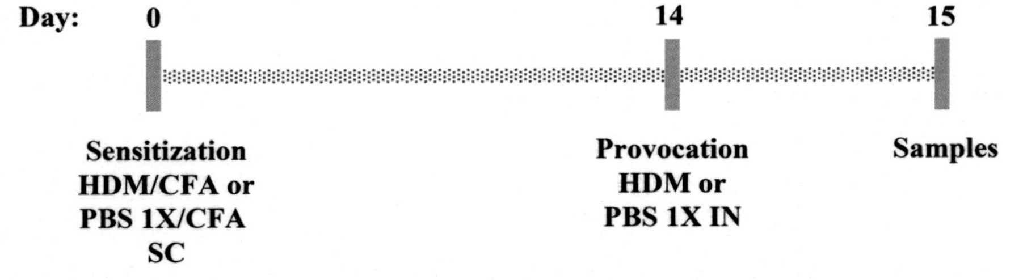

Fig. 6 Specific allergen asthma model: house dust mite (HDM)-induced severe asthma model. *SC* sub-cutaneo, *IN* intranasal, *HDM* house dust mite, *CFA* complete Freund's adjuvant, *PBS* phosphate-buffered saline

4. Grow the mold cultures for 3 weeks at 27 °C in flasks containing 250 ml of Czapek's medium.
5. Harvest the mold pellicles and homogenize in 0.4% NH_4HCO_3 + polyvinyl polypyrrolidone with an ultra-thurax.
6. Agitate the homogenates for 3 h at 4 °C.
7. Centrifuge the extracts twice, 30 min at 20,000 × *g*, dialyze against 1× PBS, and store at −20 °C in 50% glycerol.
8. Perform the experiments with 6–8 weeks old female BALB/c mice.
9. To induce the allergic lung inflammation, sensitize by intraperitoneal immunization with 2×10^6 spores emulsified in aluminum hydroxide on day 0 and day 7 (Fig. 7).
10. In control mice, use 1× PBS aluminum hydroxide.
11. On day 13, 14, and 15, challenge the mice intranasal with 2×10^5 spores.
12. Give an intraperitoneal dose of ketamine (75 mg/kg) + Medetomidine (1 mg/kg) mixture.
13. Perform the allergen deposition intranasal with 100 μl of the spore solution.
14. On day 17 (48 h after the last allergen exposure), proceed to the collection of the samples.
15. Anesthetize the animal with the same mixture described in **step 12**.
16. With the anesthetized animal, proceed to the necropsy, setting the animal in *decubitus supino* position for obtaining the samples (blood, BALF, tissues, and/or hematopoietic organs).

3.3 Tobacco-Induced Asthma Model

1. Perform the experiments with 6–8 weeks old female BALB/c mice.
2. Divide a population of 8 mice into 2 groups, one treat with ovalbumin (OVA) (sensitized) and the other with 1× PBS (control) (Fig. 8).
3. Sensitize mice with 10 μg OVA adsorbed in 250 μg of aluminum hydroxide 0.2 ml by intraperitoneal injection on day 0, 5,

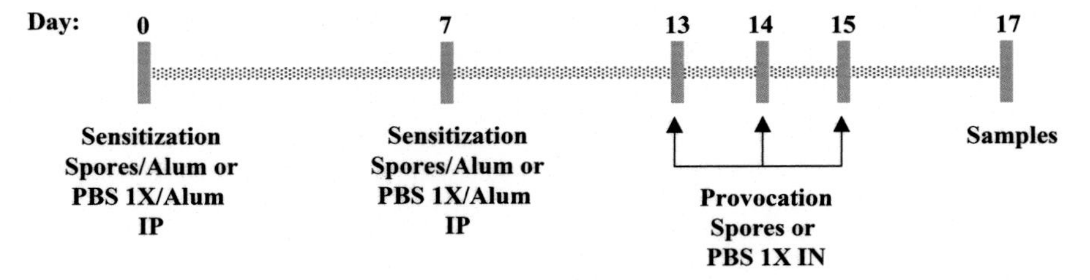

Fig. 7 Specific allergen asthma model: molds. *IP* intraperitoneal, *IN* intranasal, *PBS* phosphate-buffered saline, *Alum* aluminum hydroxide

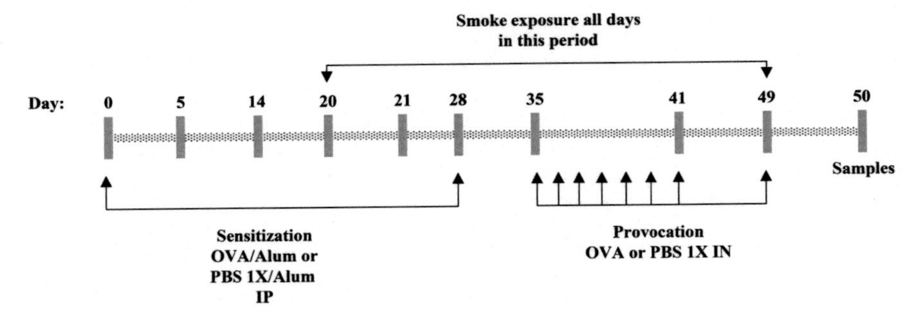

Fig. 8 Tobacco-induced asthma model. *IP* intraperitoneal, *N* Nebulization, *OVA* ovalbumin, *Alum* aluminum hydroxide, *PBS* phosphate-buffered saline

14, 21, and 28 in all mice except the controls that are treated with 1× PBS.

4. One week after the final injection, nebulize the mice with 2% OVA for 7 consecutive days from day 35 to 41, and then again nebulize with 2% OVA on day 49.
5. Nebulize the control mice with 1× PBS.
6. Divide each group in two subgroups. Expose two mice to the tobacco smoke and the other two mice not (1× PBS).
7. Expose the mice to tobacco smoke produced by a cigarette in a Plexiglas chamber (16 × 25 × 16 cm) with an inlet for pressurized air (air: smoke = 3:1).
8. Expose mice from the day 20 to day 49 (30 days), with or without smoke exposure at same times.
9. On the first day (day 20), administer the smoke of a cigarette for 10 min, and on the second day the smoke of two cigarettes.
10. Increase gradually the amount of cigarette smoke daily, one cigarette per day for 5 days.
11. The interval between smoke exposures is 15 min.
12. After the fifth day, expose the animals to five cigarettes per day from day 25 to day 49.
13. Proceed to the necropsy of the mice the following day (day 50).

Chronic asthma models can also be developed using LPS. The experiments can be performed as well with 6–8 weeks old female BALB/c mice. In these models, the mice are exposed to multiple intratracheal instillations of E. coli LPS with a delivered dose of (5 μg/mouse) equivalent to smoking approximately 25 cigarettes. In contrast to smoking models, the amount of insult added is known. The chronic inflammation is induced by giving mice a non-surgical intratracheal instillation of LPS (2 doses per week) for a total of 12 weeks.

Other interesting model is the intrinsic asthma model. The hyperreactivity is demonstrated to have a marked broncoconstriccion induced by methacholine. A widely used for the study of intrinsic asthma model is the mouse strain A/J. This strain presents bronchial hyperreactivity without treatment, handling and independently to the allergen employed.

4 Notes

1. Mice can be bred from established colonies or bought in laboratories specialized in breeding animals for research (e.g., Charles Rivers Lab). If you buy from an outside laboratory, it is advisable to acquire animals with 6 weeks of life, because, once received must remain in quarantine for 8 days as an adaptation to the recipient animal house. Thus begin experiments with them at 7 weeks. Start with mice of the same age in all experiments. Calculate the number of animals according to the specific requirements of your experiment.
2. In addition to the BALB/c strain, also can use the C57BL strain. In our studies, the BALB/c strain provides a better and stronger response to the treatment. To see the differences between the two strains at the level of experimentation in asthma, *see* Chapter 15 of this book.
3. The animals are kept in standard housing conditions. Keep them under sterile conditions, in cages (no more than 6 mice per cage) placed on racks thermostatic to 37 °C and protected with filter caps. The animals are maintained under a 12-h light-dark cycle, with *ad libitum* access to water and standard laboratory food. The animal headling is done in laminar flow hood to maintain sterility.
4. All experimental procedures must conform to international standards of animal welfare and approved by the Animal Experimentation Ethics Committee of the work institution.
5. The aluminum hydroxide acts as adjuvant allowing a slower and prolonged release of antigen (OVA in this case).
6. It has to achieve enough level of sedation to work properly with animals, but not overdo it to avoid staying too long sedated and thus facilitate their post-sedation recovery.
7. Due to the small size of mice, the volumes of the inoculated anesthetic mixture should be low. To inoculate a larger volume and thus work more safely, you should dilute the anesthetic mixture ½ in 1× PBS.
8. The anesthetic mixture is inoculated in the lower right part of the animal. For this maneuver, put the head of the animal

down to avoid visceral perforation during the process. Also, employ short needles for this step.

9. Once the anesthetic mixture is inoculated, the animal is placed in the cage until sedation. The level of animal consciousness can be assessed by auscultation of the palmar and ocular reflex.
10. For controlling the sedation level, the animal should recover in a period not exceeding 1 h. This time is more than enough to perform the maneuvers at that point.
11. To follow lung disease in mice, it can determine the respiratory frequency by counting the number of breaths per min, both pre- and post-provocation.
12. Increase the volume of dilute anesthetic mixture or use undiluted, to ensure a deeper level of sedation, if desired in this step.
13. The Clinical Score establish a scale of 0–5 as a function of the signs observed as described below:
    - Score 0: No symptoms.
    - Score 1: Scratching head and snout.
    - Score 2: Periocular-peribuccal edema, piloerection, decreased activity, and increased to a respiratory rate.
    - Score 3: Breath sounds, increased work of breathing, cyanosis peribuccal and tail.
    - Score 4: Absence of activity in response to stimulation, tremors, or seizures.
    - Score 5: Dead.
14. Due to its long duration, the chronic model may be necessary to analyze an in vivo sample taken at an intermediate point. The sample will be mainly blood that can be obtained from several extraction points. Without anesthesia, we can get it from the dorsal part of the rear leg, pedea or saphenous, with anesthesia from the tail vein or carotid [18].

Acknowledgments

This work was supported by grants of the Junta de Castilla y León ref. GRS1047/A/14 and GRS1189/A/15.

References

1. Kim HY, DeKruyff RH, Umetsu DT (2010) The many paths to asthma: phenotype shaped by innate and adaptive immunity. Nat Inmununol 7:577–584
2. Nials AT, Uddin S (2008) Mouse models of allergic asthma: acute and chronic allergen challenge. Dis Model Mech 1:213–220
3. McMillan SJ, Xanthou G, Lloyd CM et al (2005) Therapeutic administration of budesonide ameliorates allergen-induced airway remodelling. Clin Exp Allergy 35:388–396
4. Tomkinson A, Duez C, Cieslewicz G et al (2001) A murine IL-4 receptor antagonist that

inhibits IL-4- and IL-13-induced responses prevents antigen-induced airway hyperresponsiveness. J Inmunol 166:5792–5800
5. Henderson WR, Lewis DB, Albert RK et al (1996) The importance of leukotrienes in airway inflammation in a mouse model if asthma. J Exp Med 148:1483–1494
6. Temelkovski J, Hogan SP, Shepherd DP et al (1998) An improve murine model of asthma. Thorax 53:849–856
7. Henderson WR, Tang L, Chu SJ et al (2002) A role for cysteinyl leukotrienes in airway remodelling in a mouse asthma model. Am J Repir Crit Care Med 164:108–116
8. Fraga-Iriso R, Núñez-Naveira L, Brienza NS et al (2009) Desarrollo de un modelo murino de inflamacion y remodelacion de vias respiratorias en asma experimental. Arch Bronconeumol 45(9):422–428
9. Zosky GR, Sly PD (2007) Animal models of asthma. Clin Exp Allergy 37:973–988
10. Kumar RK, Herbert C, Foster PS et al (2008) The classical ovalbumin challenge model of asthma in mice. Curr Drug Targets 9:485–494
11. Lloyd CM (2007) Building better mouse models of asthma. Curr Allergy Asthma Rep 7:231–236
12. Cates EC, Fattouh R, Wattie J et al (2004) Intranasal exposure of mice to house dust mite elicits allergic airway inflammation via a GM-CSF-mediated mechanism. J Immunol 173:6384–6392
13. Conejero L, Higaki Y, Baeza ML et al (2007) Pollen-induced airway inflammation, hyper-responsiveness and apoptosis in a murine model of allergy. Clin Exp Allergy 37:331–338
14. Havaux X, Zeine A, Dits A et al (2004) A new mouse model of lung allergy induced by the spores of Alternaria alternata and Cladosporium herbarum molds. Clin Exp Immunol 139: 179–188
15. Maes T, Provoost S, Lanckacker EA et al (2010) Mouse models to unravel the role of inhaled pollutants on allergic sensitization and airway inflammation. Respir Res 11:7
16. Kim DY, Kwon EY, Hong GU et al (2011) Cigarette smoke exacerbates mouse allergic asthma through Smad proteins expressed in mast cells. Respir Res 12:49
17. Hadeiba H, Corry DB, Locksley RM (2000) Baseline airway hyperreactivity in A/J mice is not mediated by cells of the adaptive immune system. J Immunol 164(9):4933–4940
18. Hoff J (2000) Methods of blood collection in the mouse. Lab Anim 29:10

Chapter 17

Mouse Models Applied to the Research of Pharmacological Treatments in Asthma

Fernando Marqués-García and Elena Marcos-Vadillo

Abstract

Models developed for the study of asthma mechanisms can be used to investigate new compounds with pharmacological activity against this disease. The increasing number of compounds requires a preclinical evaluation before starting the application in humans. Preclinical evaluation in animal models reduces the number of clinical trials positively impacting in the cost and in safety. In this chapter, three protocols for the study of drugs are shown: a model to investigate corticoids as a classical treatment of asthma; a protocol to test the effects of retinoic acid (RA) on asthma; and a mouse model to test new therapies in asthma as monoclonal antibodies.

Key words Animal model, Antibody, Corticoids, Monoclonal, Mouse

1 Introduction

In predisposed individuals, initial exposure to the allergen leads to the activation of allergen-specific Th2 cells and IgE synthesis (allergic sensitization). Subsequent exposure to the allergen causes inflammatory cell recruitment, activation, and mediator release (histamine, leukotrienes, and cytokines). According to this, potential therapeutic interventions are proposed for the treatment of allergic diseases. Current treatments are primarily focused on pharmacological intervention to decrease the symptoms of the disease [1]. Among them the most used are corticosteroids, which are very effective drugs for disease control [2]. Although these treatments are highly effective for most individuals, many patients must take these drugs lifetime.

Mouse models can also be used to investigate the effect of other potential compounds. This is the case of all-trans-retinoic acid (ATRA). The ATRA derived from vitamin A in the diet [3] regulates cell growth, differentiation, and matrix formation in various cell types and plays a major role in the maintenance of a normal epithelial mucociliar phenotype [4].

María Isidoro-García (ed.), *Molecular Genetics of Asthma*, Methods in Molecular Biology, vol. 1434,
DOI 10.1007/978-1-4939-3652-6_17, © Springer Science+Business Media New York 2016

Besides its modulatory role on epithelial cells, RA suppresses the differentiation of eosinophil and basophil from bone marrow precursors [5], most likely by controlling the IL-5 receptor expression on their surface [6], and by preventing mitogen-induced proliferation and IL-4-dependent IgE production [7]. It has been suggested a potential therapeutic effect of RA in asthma.

Corticosteroid treatment may not be effective in some patients or resistances to inhaled corticoids may appear [8]. The therapeutic use of monoclonal antibodies presents an opportunity for the treatment of these patients. Omalizumab was the first therapeutic antibody approved by FDA (U.S. Food and Drug Administration) for treatment of asthma and the first approved therapy designed to target immunoglobulin E (IgE) [9]. Today, several monoclonal antibodies for treating asthma are directed against different target [10].

Another strategy is the allergen-specific immunotherapy (SIT) that modifies the allergen-specific immune response and the course of disease [11]. Allergen-specific immunotherapy is based on repeated administration of disease-causing allergens in order to modify the allergen-specific immune response to reach tolerance [12]. In this chapter, we describe mouse models based on the experimental study of different treatments for asthma.

2 Materials

2.1 Animals

1. 6–8 Weeks old female BALB/c mice (Charles Rivers Lab).

2.2 Common Materials

1. 1× PBS (Phosphate-Buffered Saline): 1.37 mM NaCl, 2.6 mM KCl, 10 mM Na_2HPO_4, 1.8 mM KH_2PO_4. Adjust the pH to 7.4 with HCl.
2. Tuberculin syringes (BD Biosciences).
3. Needle 20G short (BD Biosciences).
4. Ketamine chlorhydrate (Imalgene 1000, Merial).
5. Medetomidine hydrochloride (Medetor, Virbac).
6. 100 μl micropipette and appropriate tips (Thermo-Fischer).

2.3 Corticoid Model

1. Dexamethasone (Sigma-Aldrich).
2. Ovalbumin (OVA) (Sigma-Aldrich).
3. Aluminum Hydroxide (Sigma-Aldrich).

2.3.1 House Dust Mite (HDM)-Induced Asthma Model

1. Budesonide (Sigma-Aldrich).
2. House Dust Mite (HDM) Extract (Allergopharma).

2.3.2 House Dust Mite (HDM)-Induced Severe Asthma Model

1. Budesonide (Sigma-Aldrich).
2. House Dust Mite (HDM) Extract (Allergopharma).
3. Freund's adjuvant: 1 ml contains 1 mg of *Mycobacterium tuberculosis* (H37Ra, ATCC 25177), heat killed and dried, 0.85 ml paraffin oil, and 0.15 ml mannide monooleate (Sigma-Aldrich).

2.4 Asthma Model for Acid All-Trans-Retinoic (ATRA) Treatment

1. All-Trans-Retinoic Acid (ATRA) (Sigma-Aldrich).
2. DPLC (dilauroylphosphatidylcholine) (Sigma-Aldrich).
3. Butanol.
4. Isopropanol.

2.5 Asthma Model for Treatment with Monoclonal Antibodies

1. Ovalbumin (OVA) (Sigma-Aldrich).
2. Aluminum Hydroxide (Sigma-Aldrich).
3. Anti-mouse–IL-5 monoclonal.
4. Isotype control monoclonal antibody.

3 Methods

3.1 Corticoid Models

3.1.1 Acute Model of Asthma for Corticoid Treatment

In this chapter, we will review a protocol of acute asthma in mouse to be treated with dexamethasone.

1. Perform the experiments with 6–8 weeks old female BALB/c mice.
2. Animals must be kept confined and handled to meet the current ethics rules of handling animals for experimentation.
3. It starts with 6–8 weeks old female BALB/c mice, for example eight mice. Divide the animals into two groups (Fig. 1):
   - Group 1: Four mice that will not be injected with OVA Intraperitoneal/Intranasal (in place 1× PBS will be given).

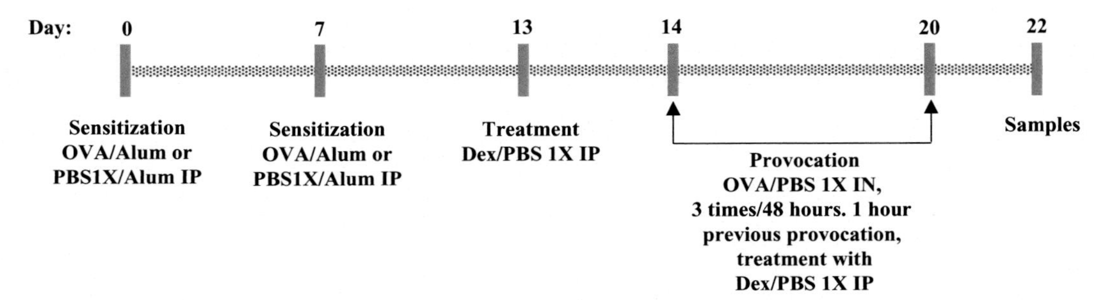

Fig. 1 Corticoids: acute mouse model. *IP* intraperitoneal, *IN* intranasal, *Dex* dexamethasone, *PBS* phosphate-buffered saline, *OVA* ovalbumin, *Alumn* aluminum hydroxide

- Group 2: Four mice in which OVA Intraperitoneal/Intranasal will be administered.

4. On day 0 (start of experiment), inject the group 2 by intraperitoneal route, with 20 μg of OVA emulsified in 0.2 ml 1× PBS containing 2 mg of aluminum hydroxide (these doses are administered to each mouse).
5. In the controls, group 1, inject with 0.2 ml of 1× PBS containing 2 mg of aluminum hydroxide.
6. Keep the animals stabled for 1 week (7 days) for the OVA hyper-sensitization to occur.
7. On day 7, administer the second OVA dose to reinforce the hyper-sensitization. For this, the OVA intraperitoneal injection is repeated as in **step 4**.
8. Keep the mice housed another 7 days.
9. On day 13 of the experiment, split each of the mice populations in two, as described below (*see* **Note 1**):
   - Group 1: Without OVA
     (a) Two mice with 1× PBS (control).
     (b) Two mice with dexamethasone (control treated with dexamethasone).
   - Group 2: With OVA.
     (a) Two mice with 1× PBS (asthma without treatment).
     (b) Two mice with dexamethasone (asthma with treatment).
10. That same day, carry out the first intraperitoneal injection of corticosteroid (dexamethasone) or 1× PBS, based on the designed groups of mice.
11. Use dexamethasone at a concentration of 3 mg/kg. Inject each mouse with 60 μg of dexamethasone dissolved into 0.2 ml of 1× PBS (*see* **Note 2**).

    In summary, we have the following treatment groups:
   - Intraperitoneal/Intranasal 1× PBS + 1× PBS intraperitoneal.
   - Intraperitoneal/Intranasal 1× PBS + dexamethasone intraperitoneal.
   - Intraperitoneal/Intranasal OVA + 1× PBS intraperitoneal.
   - Intraperitoneal/Intranasal OVA + dexamethasone intraperitoneal.
12. The next day, day 14, perform other injection with intraperitoneal dexamethasone or 1× PBS, 1 h before intranasal provocation.

13. An hour late, perform the first intranasal provocation with OVA or 1× PBS.
14. First, anesthetize the animal. Administer an intraperitoneal dose of Ketamine (75 mg/kg) + Medetomidine (1 mg/kg). Wait for a complete sedation of the animal, before starting to handle (*see* **Note 3**).
15. In sedated animal, standing upright with his nose up, give 50 μl of 1× PBS to the control group or 50 μl of OVA (0.1 % in 1× PBS) to the experimental group. Administer the mixture drop wise on the snout, favoring inoculation during the inspiratory process of the animal.
16. Keep the animal in that position for 1–2 min. Return the animal to the cage, and place it horizontally on its side for favor ventilation.
17. Repeat **steps 12–17**, three times with a frequency of 2 days (48 h) between a provocation and the next. In this protocol, it corresponds to day 16 (second provocation), 18 (third provocation), and 20 (fourth provocation).
18. On day 22, proceed to the collection of samples, 2 days after the last challenge (48 h).
19. Anesthetize the animal with the same mixture described in **step 14**.
20. With the anesthetized animal, proceed to the necropsy. The mouse is placed in ***decubitus supino***, fixing the animal to a surface by the legs.
21. For obtaining samples:
    - Blood is obtained by cardiac puncture at the ventricular level. This puncture serves as blood gathering and euthanasia. Cellular and soluble factors can be analyzed.
    - BALF (Bronchoalveolar Lavage) can be obtaining by cannulating the trachea, injecting 1 ml of 1× PBS, two times, and collect. BALF is used for cellular and soluble factor studies.
    - Collect the organs of interest for molecular biology as well as histological studies.
    - Collection of hematopoietic organs can be performed for in vitro cell culture.

3.1.2 Chronic Model of Asthma for Corticoid Treatment

1. Perform the experiments with 6–8 weeks old female BALB/c mice (Fig. 2).
2. Animals must be kept confined and handled to meet the current ethics rules of handling animals for experimentation.
3. It starts with a population of 6–8 weeks old female BALB/c mice. Divide the animals into two groups:

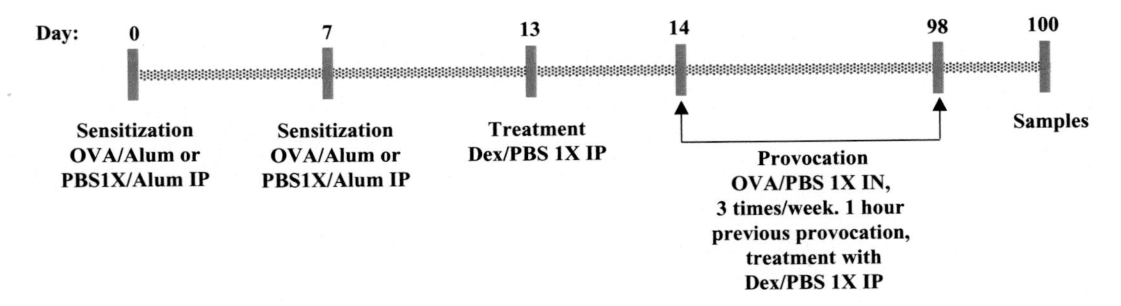

Fig. 2 Corticoids: chronic mouse model. *IP* intraperitoneal, *IN* intranasal, *Dex* dexamethasone, *PBS* phosphate-buffered saline, *OVA* ovalbumin, *Alum* aluminum hydroxide

- Group 1: Four mice that will not be treated with OVA Intraperitoneal/Intranasal (in place 1× PBS will be given).
- Group 2: Four mice that will be treated with OVA Intraperitoneal/Intranasal.

4. On day 0 (start of experiment), inject the animals of the group 2, by intraperitoneal route, with 20 μg of OVA emulsified in 0.2 ml 1× PBS containing 2 mg of aluminum hydroxide (these doses are administered to each mouse).

   Inject the animals of the control group, group 1, with 0.2 ml of 1× PBS containing 2 mg of aluminum hydroxide.
5. Keep the animals stabled for 1 week (7 days) for the reaction of hyper-sensitization to occur with OVA.
6. On day 7, administer the second dose of OVA to reinforce the hyper-sensitization. For this, the intraperitoneal injection is repeated as in **step 4**.
7. Keep the mice housed for another 7 days.
8. On day 13 of the experiment, divide each population of mice into two groups, as described below:
   - Group 1: Without OVA.
     (a) Two mice with 1× PBS (control).
     (b) Two mice with dexamethasone (control treated with dexamethasone).
   - Group 2: With OVA.
     (a) Two mice with 1× PBS (asthma without treatment).
     (b) Two mice with dexamethasone (asthma with treatment).
9. That same day carry out the first intraperitoneal injection of corticosteroid (dexamethasone) or 1× PBS, based on the groups of mice performed.

10. Use dexamethasone at a concentration of 3 mg/kg, thus injecting mice with 5 μg of dexamethasone into 0.2 ml of 1× PBS.

    In summary, we have the following treatment groups:

    - Intraperitoneal/Intranasal 1× PBS + 1× PBS intraperitoneal.
    - Intraperitoneal/Intranasal 1× PBS + dexamethasone intraperitoneal.
    - Intraperitoneal/Intranasal OVA + 1× PBS intraperitoneal.
    - Intraperitoneal/Intranasal OVA 1× + dexamethasone intraperitoneal.

11. The next day, day 14, perform other intraperitoneal injection of 1× PBS or dexamethasone 1 h before the intranasal provocation.
12. An hour later, perform the first intranasal provocation with OVA or 1× PBS.
13. First anesthetize the animal. Administer an intraperitoneal dose of Ketamine (75 mg/kg) + Medetomidine (1 mg/kg). Wait for a complete sedation of the animal before starting.
14. Stand upright the sedated animal, with his nose up and give 50 μl of 1× PBS to the control group or 50 μl of OVA (0.1% in 1× PBS) to the experimental group. Administer drop wise to the snout, favoring inoculation during the inspiratory process of the animal.
15. Keep the animal in that position for 1–2 min.
16. Return the animal to the cage, putting horizontally on its side, for favoring his ventilation.
17. In this chronic model, it is interesting to monitor the clinical course of the mice. Use a classification or Clinical Score that assigns a numerical value increasing with the severity of the response according to the clinical signs developed.
18. Repeat the procedure described in **steps 12–17**, three times per week for 12 weeks (3 months).
19. Observe and record the signs in mice during this period.
20. After that time, proceed to the collection of samples.
21. Perform the necropsy of the animals, 2 days (48 h) after the last provocation.
22. Anesthetize the animal with the same mixture described in paragraph 14.
23. With the anesthetized animal, proceed to the necropsy. The mouse is placed in *decubitus supino* fixing it by the legs to a surface.

24. For obtaining samples:
    - Blood is obtained by cardiac puncture at the ventricular level. This puncture serves as blood gathering and euthanasia. Cellular and soluble factors can be analyzed.
    - BALF (Bronchoalveolar Lavage) can be obtained by cannulating the trachea, injecting 1 ml of 1× PBS, two times, and collect. BALF is used for cellular and soluble factors studies.
    - Collect the organs of interest for molecular biology as well as histological studies.
    - Collection of hematopoietic organs can be performed for in vitro cell culture.

3.1.3 House Dust Mite (HDM)-Induced Asthma Model for Corticoid Treatment

All allergen-specific models, as well as those described in Chapter 16 of this book, may be used for corticosteroid treatment studies.

1. For the acute model, dissolve HDM in 1× PBS at a concentration of 10 μg of extract (plus 0.5 mg aluminum hydroxide) in 50 μl of 1× PBS per mouse in the experimental group. In control mice, give 50 μl of 1× PBS with 0.5 mg of aluminum hydroxide.
2. Perform the experiments with 6–8 weeks old female BALB/c mice.
3. Start for example from a population of 8 mice. Divide the animals into two groups:
    - Group 1: Four mice without Intraperitoneal HDM (in place will be given 1× PBS).
    - Group 2: Four mice with HDM Intraperitoneal
4. The first day of the experiment (day 0), perform the first administration of HDM extract or 1× PBS (plus adjuvant for both), depending on the group (sensitization phase). In this case, it is administered by intraperitoneal (Fig. 3).

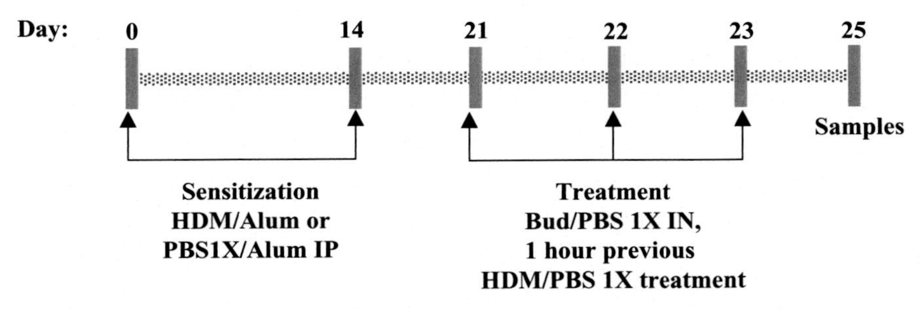

Fig. 3 Corticoids: house dust mite (HDM)-induced asthma model. *IP* intraperitoneal, *IN* intranasal, *Bud* budesonide, *PBS* phosphate-buffered saline, *HDM* house dust mite, *Alum* aluminum hydroxide

5. Keep the animals housed for 14 days.
6. On day 14, repeat **step 3**.
7. Maintain mice housed until day 21 of the experiment.
8. Prepare HDM at a concentration of 10 μg of extract in 50 μl of 1× PBS without aluminum hydroxide.
9. On day 21, perform the first administration of corticosteroids.
10. Use budesonide (0.3 mg/kg in 1× PBS), administer intranasal.
11. Furthermore, perform the first challenge by administering the allergen prepared in paragraph eight for intranasal route.
12. Administer an intraperitoneal dose of ketamine mixture (75 mg/kg) + Medetomidine (1 mg/kg), achieving a smooth sedation.
13. Administer corticosteroid 1 h prior to sensitization to HDM.
14. Perform allergen deposition intranasal.
15. Repeat **steps 10–13**, days 22 and 23 of the experiment.
16. On day 25 (48 h after the last allergen exposure and the last dose of corticosteroid), proceed to the samples collection.
17. Anesthetize the animal with the same mixture described in **step 12**.
18. With the anesthetized animal, proceed to the necropsy, setting the animal in *decubitus supino* position.
19. Obtain samples (blood, BALF, tissues, and/or hematopoietic organs).

3.1.4 House Dust Mite (HDM)-Induced Severe Asthma Model for Corticoid Treatment

1. Dissolve HDM in 1× PBS at a concentration of 100 μg extract more complete Freund's adjuvant in 50 μl of 1× PBS per mouse in the experimental group. In control mice, give 50 μl of 1× PBS plus adjuvant (Fig. 4).
2. Perform the experiments with 6–8 weeks old female BALB/c mice.

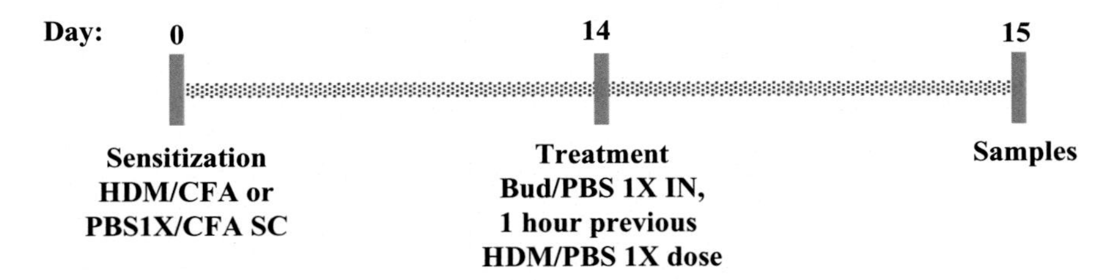

Fig. 4 Corticoids: house dust mite (HDM)-induced severe asthma model. *SC* subcutaneous, *IN* intranasal, *Bud* budesonide, *PBS* phosphate-buffered saline, *HDM* house dust mite, *CFA* complete Freund's adjuvant

3. On first day of the experiment (day 0), perform the first administration of HDM extract or 1× PBS (plus adjuvant for both), depending on the group (HDM or control) (sensitization phase). In this case, administer via subcutaneous.
4. Keep the mice housed for 14 days.
5. Prepare the HDM at a concentration of 100 μg of extract in 50 μl of 1× PBS, without adjuvant.
6. On day 14 day, perform corticosteroid administration in single dose.
7. Use budesonide as corticosteroid (0.3 mg/kg in 1× PBS), intranasal.
8. Furthermore, the provocation model is performed by administering the allergen prepared in **step 5**, via intranasal.
9. Administer an intraperitoneal dose of ketamine mixture (75 mg/kg) + Medetomidine (1 mg/kg), achieving a mild sedation.
10. Deposit the corticosteroid 1 h before challenge.
11. The allergen deposition is performed intranasal.
12. On day 15 (24 h after challenge), proceed to the sample collection.
13. Anesthetize the animal with the same mixture described in **step 9**.
14. With the anesthetized animal, proceed to the necropsy, setting the animal in *decubitus supino* position.
15. Obtain the samples as needed (blood, BALF, tissues, and/or hematopoietic organs).

3.2 Asthma Model for All-Trans-Retinoic Acid (ATRA) Treatment

In this section, we will focus on a protocol design to study the effects of a pretreatment with ATRA over an acute model of asthma (*see* **Note 4**).

1. Initially, prepare a solution of 20 μg of OVA, emulsified in 0.2 ml of 1× PBS containing 2 mg of aluminum hydroxide (this dose is administered per mouse).
2. Secondly, prepare ATRA. ATRA is a fat-soluble molecule to be encapsulated in a vehicle to promote their distribution in the animal.
3. Prepare the liposome particle DPLC. Dissolve 100 mg DPLC in 1 ml of butanol. This volume will be split in two tubes, one will be supplemented by ATRA, the other not.
4. Moreover, dissolve ATRA in isopropanol at a concentration of 100 mg/kg. 0.2 ml 1× PBS of ATRA (10%, wt:wt) per animal and day of dosing (*see* **Note 5**).
5. Mix the solution of DPLC-ATRA with a ratio of 4 parts ATRA to 1 part DPLC. Prepare aliquots depending on the number of mice and days of inoculation.

6. Prepare all solutions with butanol or isopropanol, leave in a hood until complete evaporation of the alcohol. Once the dried pellet is obtained, store at −80 °C until use.
7. Perform the experiments with 6–8 weeks old female BALB/c mice (*see* **Note 6**).

   The mice will be divided into two experimental groups (Fig. 5):

   - Control group without ATRA (for example, 4 mice).
   - Group treated with ATRA (for example, 4 mice).
8. One day before the sensitization of mice (day-1), administer the first dose of liposome-ATRA to the treated group, and the first dose of empty liposome to the control group.
9. Dissolve each aliquot (both liposome-ATRA as empty liposome) in 0.3 ml of 1× PBS.
10. Inoculate intraperitoneal.
11. One day later (day 0), repeat **steps 8** and **9** an hour before the sensitization.
12. Next, divide the groups as follows:
   - Control without ATRA:
     (a) Two mice with 1× PBS (control).
     (b) Two mice with ovalbumin (OVA).
   - Group with ATRA:
     (a) Two mice with 1× PBS.
     (b) Two mice with ovalbumin (OVA).

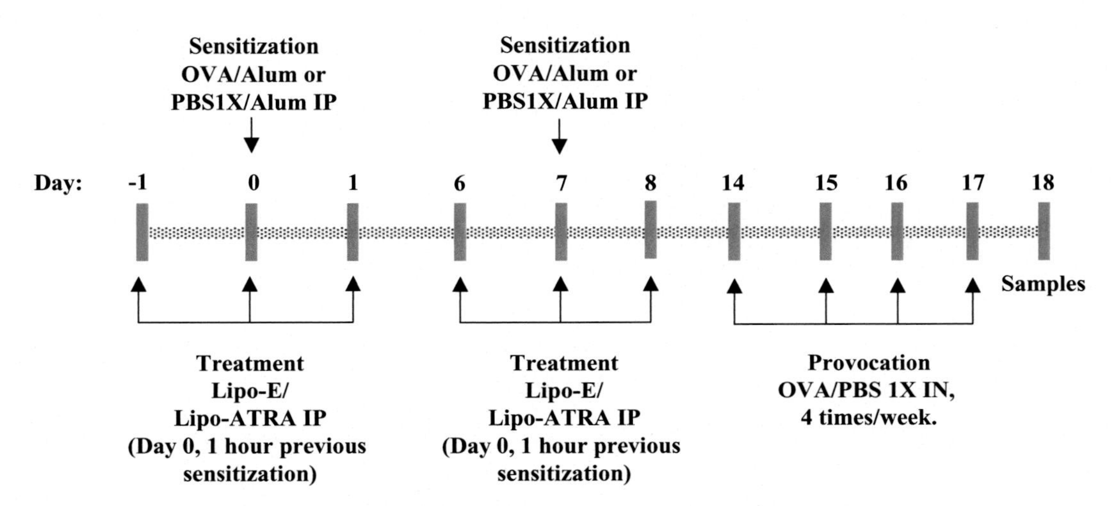

Fig. 5 All-trans-retinoic acid model (ATRA). *IP* intraperitoneal, *IN* intranasal, *PBS* phosphate-buffered saline, *OVA* ovalbumin, *Alum* aluminum hydroxide, *Lipo-ATRA* liposome-all-trans-retinoic acid, *Lipo-E* liposome-empty

13. One hour after injection of ATRA, carry out the sensitization. Inoculate intraperitoneal 0.2 ml of the OVA solution prepared in **step 3**. Inoculate 1× PBS to the controls.
14. The next day (day 1), administer another dose of liposome-ATRA (**steps 8–9**).
15. On day 6 (1 day before the second sensitization), administer the fourth dose of liposome-ATRA by repeating the **steps 8–9**.
16. On day 7, administer the fifth dose of liposome-ATRA 1 h before the sensitization repeating the **steps 8–9**.
17. One hour after ATRA treatment, repeat **step 13**.
18. On day 8, repeat the **steps 8–9** for the administration of the sixth dose of liposome-ATRA.
19. On day 14, begin to make provocations.

Perform the first intranasal provocation with OVA or 1× PBS according to the group.

20. Anesthetize the animal. To obtain a mild sedation, administer an intraperitoneal dose of Ketamine (75 mg/kg) + Medetomidine (1 mg/kg). Wait for a complete sedation of the animal before you start.
21. Stand upright the sedated animal, with his nose up, and administer 50 μl of 1× PBS to the control group or 50 μl of OVA (0.1 % in 1× PBS) to the experimental group. Administer the mixture drop wise to the snout, favoring the inoculation during the inspiratory process of the animal.
22. Keep the animal in that position for 1–2 min.
23. Return the animal to the cage, placing flat on its side to favor ventilation.
24. Repeat the challenge, the days 15, 16, and 17 of the experiment.
25. On day 18 (24 h after the last challenge), proceed to the sample collection.
26. Anesthetize the animal with the same mixture described in the **step 20**.
27. With the anesthetized animal, proceed to the necropsy, setting the animal in *decubitus supino* position.
28. Proceed to obtain the samples (blood, BALF, tissues, and/or hematopoietic organs) (*see* **Note 7**).

3.3 Monoclonal Antibodies

This section relies on chronic corticosteroid treatment model described in this chapter, making some modifications.

1. On day 0 (start of experiment), inject the animals (4 mice), by intraperitoneal route, with 50 μg of OVA, emulsified in 0.2 ml 1× PBS containing 2 mg of aluminum hydroxide (these doses

are administered to each mouse). Inject the controls (4 mice) with 0.2 ml of 1× PBS containing 2 mg of aluminum hydroxide (Fig. 6).

2. Perform a sensitization on day 14 of the experiment, as described in **step 1**.
3. On day 21 of the experiment, divide mice in two groups (for example, starting 4 per group):
   - Group 1: Without OVA.
     (a) Two mice with isotype control monoclonal antibody (0.5 mg per injection).
     (b) Two mice with anti-mouse-IL-5 monoclonal (0.5 mg per injection).
   - Group 2: With OVA.
     (a) Two mice with isotype control monoclonal antibody (0.5 mg per injection).
     (b) Two mice with anti-mouse-IL-5 monoclonal (0.5 mg per injection).
4. Challenge the animals with intranasal OVA 3 days/week for 6 weeks.
5. During the last 2 weeks of treatment with OVA, administer monoclonal antibody doses by intraperitoneal route, days 50, 55, 59, and 63.
6. On day 65, 48 h after the last challenge proceed to sample collection.

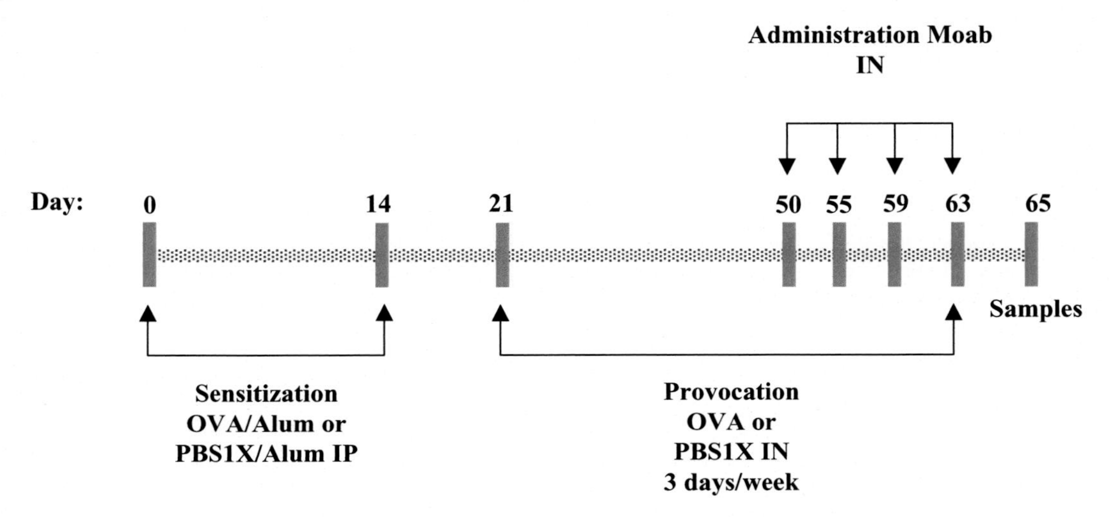

Fig. 6 Monoclonal antibodies. *IP* intraperitoneal, *IN* intranasal, *PBS* phosphate-buffered saline, *OVA* ovalbumin, *Alum* aluminum hydroxide, *Moab* monoclonal antibody

4 Notes

1. Six mice can be stored per cage. Keep mice in a cage for group 1 and group 2 in another. It is necessary to differentiate in each cage mice treated with dexamethasone from treated with 1× PBS. Optionally, mark the animals with an incision in the ear with an ear tag. Calculate the number of animals according to the requirements of the experiment.
2. This dose of dexamethasone is calculated for an average weight of 20 g per mice.
3. It is advisable to inject the anesthetic in one side of the animal body and dexamethasone in the other, avoiding overloading an area with an excessive number of inoculations.
4. The same protocols described in the treatment sections with OVA (chronic and acute) can be applied to the studies of the effect of ATRA in asthma, replacing the intraperitoneal injection of corticosteroid for the injection of ATRA.
5. The ATRA is a photosensitive molecule, which must be protected from light when it is lyophilized and when it is in solution. Store at −20 °C.
6. It is necessary to know the amount of vitamin A that the animal is consuming in the diet.
7. To assess levels of ATRA, the HPLC methodology can be used.

Acknowledgments

This work was supported by grants of the Junta de Castilla y León ref. GRS1047/A/14 and GRS1189/A/15.

References

1. Holgate ST, Broided D (2003) New targets for allergic rhinitis-a disease of civilization. Nat Rev Drug Discov 2:902–914
2. Barnes PJ, Pedersen S, Busse WW (1998) Efficacy and safety of inhaled corticosteroids. Am J Respir Crit Care Med 157:1–53
3. Marielle M, Ruffie C, Periquet B et al (2007) Liposomal retinoic acids modulate asthma manifestations in mice. J Nutr Dis 0022-3166/07:2731-2736
4. Biesalski HK (2003) The significance of vitamin A for the development and function of the lung. Forum Nutr 6:37–40
5. Denburg JA, Sehmi R, Upham J et al (1999) Regulation of IL-5 and IL-5 receptor expression in the bone marrow of allergic asthmatics. Int Arch Allergy Immunol 118:101–103
6. Denburg JA, Sehmi R, Upham J (2001) Regulation of IL-5 receptor on eosinophil progenitors in allergic inflammation: role of retinoic acid. Int Arch Allergy Immunol 124:246–248
7. Goettsch W, Hattori Y, Sharma RP (1992) Adjuvant activity of all-trans-retinoic acid in C57Bl/6 mice. Int J Immunopharmacol 14:143–150
8. Rambasek TE, Lang DM, Kavuru MS et al (2004) Omalizumab: where does it fit into current asthma management? Cleve Clin J Med 71:251–261

9. Curtiss F (2005) Selectivity and specificity are the keys to cost-effective use of Omalizumab for allergic asthma. J Manag Care Pharm 11:774–776
10. Polosa R, Casale T et al (2012) Monoclonal antibodies for chronic refractory asthma and pipeline developments. Drug Discov Today 17(11/12):591–599
11. Larche M, Akdis CA, Valenta R (2006) Immunological mechanisms of allergen-specific immunotherapy. Nat Rev Immunol 6:761–771
12. Valenta R, Campana R, Marth K et al (2012) Allergen-specific immunotherapy: from therapeutic vaccines to prophylactic approaches. J Intern Med 272:144–157

Chapter 18

Review on Pharmacogenetics and Pharmacogenomics Applied to the Study of Asthma

Almudena Sánchez-Martín, Asunción García-Sánchez, and María Isidoro-García

Abstract

Nearly one-half of asthmatic patients do not respond to the most common therapies. Evidence suggests that genetic factors may be involved in the heterogeneity in therapeutic response and adverse events to asthma therapies. We focus on the three major classes of asthma medication: β-adrenergic receptor agonist, inhaled corticosteroids, and leukotriene modifiers. Pharmacogenetics and pharmacogenomics studies have identified several candidate genes associated with drug response.

In this chapter, the main pharmacogenetic and pharmacogenomic studies in addition to the future perspectives in personalized medicine will be reviewed. The ideal treatment of asthma would be a tailored approach to health care in which adverse effects are minimized and the therapeutic benefit for an individual asthmatic is maximized leading to a more cost-effective care.

Key words Asthma, β-adrenergic, Corticosteroids genes, Leukotriene, Pharmacogenetics, Pharmacogenomics, Personal medicine

1 Introduction

Asthma is a phenotypically heterogeneous disease caused by interactions among demographic, social, environmental, and genetic factors [1, 2]. The disease is characterized by variable degrees of airflow obstruction and airway inflammation producing breathlessness and wheezing [3]. Also, it is one of the most common chronic diseases currently affecting more than 300 million people worldwide. The annual death rate from complications of asthma approximates to 200,000 people worldwide, mainly in less developed countries. It is estimated that these numbers may be increased by 20% over the next 10 years if adequate measures are not taken [4].

It is difficult to quantify the global economic burden of asthma. The indirect cost of asthma, especially its negative impact on productivity, is very high. According to The Global Asthma Report 2014, the estimated total cost to society in the United States of America

María Isidoro-García (ed.), *Molecular Genetics of Asthma*, Methods in Molecular Biology, vol. 1434, DOI 10.1007/978-1-4939-3652-6_18,

was $56 billion in 2007, or $3259 per person per year (in 2009 US dollars). The impact of these indirect costs would be diminished by improving asthma control, through improving access to good management including medicines for controlling the disease (http://www.globalasthmareport.org/). The complexity of asthma control lies in the different response pathways of drugs and the variable responses from patients.

Nearly 70–80% of patients with asthma may have high variability in the clinical response to the most common asthma medications. Among main factors that may be responsible for this variability, poor patient adherence, ineffective methods of drug delivery, incorrect drug selection are included; however, in most cases, the interindividual differences cannot be explained only by these factors, suggesting that there may be a genetic component that could play an important role [5].

The Human Genome Project is an international initiative designed to identify the sequence of the human genome. Following with the publication in 2001 of a sequence draft, investigators have focused on genomic variations among individuals [6]. These variations include single nucleotide polymorphisms (SNPs), base insertions or deletions, copy-number variations, and variable numbers of tandem repeats. SNPs are polymorphisms that affect a single base. Some of these variations could affect the quantity, timing, and function of the encoded protein, contributing to the differences in drug response among individuals [7], providing the rationale for pharmacogenetic research.

The term pharmacogenetics was coined by Vogel in 1959 [8]. Pharmacogenetics refers to the study of inherited variations in one single gene on drug response, and pharmacogenomics is the study of the role of inherited and acquired genetic variation in many different genes (genome) including DNA and RNA that determine drug response [9, 10]. The goal of pharmacogenomics is to facilitate the identification of biomarkers that can help optimize drug selection, dose, and treatment duration and prevent adverse reactions. Pharmacogenomics can provide new insight into mechanisms of drug action and contribute to development of new therapeutics [9].

2 Methods for Discovery Pharmacogenetic Variants

The approaches in pharmacogenomic studies have evolved from individual candidate gene studies for evaluation of drug pathways to genome-wide association studies (GWASs) to identify novel variations affecting drug response [10]. In addition, analysis of other data, like expression or biochemical data, may identify gene candidates.

2.1 Studies of Candidate Gene

The candidate gene association studies have been used in many pharmacogenetic studies of asthma. These approaches interrogate whether a particular allele or a set of alleles is more frequent in patients who have a better or worse drug response [11, 12]. Candidate gene studies are relatively cheap and quick to perform. Genes are selected based on their known physiological or pharmacologic effect on disease or drug response. So a previous knowledge about the function of a gene is essential to select the candidate gene to study. This is followed by identification and selecting SNPs having a functional consequence as the function of a protein. Finally, the gene variant is associated to the drug response (trait) by detecting its existence in random test subjects (cases) having the trait and the selected control subjects, which do not [12].

In pharmacogenetic studies, SNPs with minor allele frequencies (MAF) above 0.10 are generally selected for detecting genetic associations. SNPs with MAFs lower than 0.10 are less likely to report a genetic association. An exception occurs when SPNs with low MAFs causes functional effects, such as an alteration in gene expression or in the functional effects on the protein [5]. The principal advantage of this approach is that requires a relative small sample size, is cheap, simpler to perform and to interpret because the physiological or pharmacologic effects of the genes are previously well characterized. The main disadvantage is to require previous knowledge of the function of the gene concerning to drug response. If information about the gene function is limited, the selection of the gene is difficult to support.

2.2 Biological Pathway Studies

This approach analyzes multiple genes encoding several components of a drug response pathway, involving important genes for controlling steps in drug response. The strategy for selecting genetic SNPs is nearly identical to candidate gene studies [5].

2.3 Genome-Wide Association Studies (GWAS)

GWAS allow the analysis of hundreds of thousands of SNPs with the goal of relating them to a specific phenotype, a disease, or a better or worse drug response. In a GWAS, the SNPs representing regions of the genome with the most interindividual variation are probed on a DNA microarray for each individual in a set of cases and controls. Each SNP is tested independently and the significance (p-value) must be corrected for multiple hypotheses, usually using Bonferroni correction [13]. When an SNP reaches the genome-wide significance, it is a candidate for following analysis.

The main advantage of GWASs has been their excellent resolution, their statistical power to detect risk variants with moderate effects, and to be able of analyze the whole genome. On the contrary, GWASs need large sample sizes and tools for processing a very large pool of data. A huge number of statistical comparisons (hundreds of thousands) make that the threshold defining significant associations should be lower than conventional case-control studies [6, 14]. Like the candidate gene association studies, the results should be replicated in multiple independent populations.

As the price of high-throughput sequencing is decreasing, many researchers are performing exome or whole-genome sequencing to identify genetic factors of drug response. This approach is also hypothesis-independent in the identification of SNPs, detects less common mutations, and captures larger-scale information, including copy-number variants (CNVs) and structural variants [13].

2.4 RNAseq

RNA expression data from microarrays or RNAseq from drug-treated samples could serve as start point in discovering new genes involved in drug response. Using expression profiles generated from a drug-treated sample compared to control can be used to determine a molecular response to a drug. Usually, these experiments are carried out in cell lines that provide a valuable low-cost resource [15].

2.5 Expression Quantitative Trait Loci (eQTL)

eQTL are genomic loci that contribute to variation in expression levels of mRNAs or protein, usually a single gene product with a specific chromosomal location. There are cis-eQTLs that map near the gene and can act locally, and those mapped far even in another chromosome, the trans-eQTL, which can act at a distance (distant eQTLs). The integration of the analysis of DNA and RNA sequencing offers a unique view of the genetic causes of variation in the transcriptome. The variants of eQTL and the consequent variation of expression level can add a functional dimension to our knowledge, helping the understanding of the mechanisms leading this variation [16].

2.6 Pathway Discovery

Once the candidate gene is identified, analyzing genetic networks, cascades, and pathways can help to discover other possible candidates that affect the drug action. The information on the gene's network or pathway could serve to limit or generate new hypotheses in the pharmacogenomics of the drug.

3 Asthma Therapy: Pharmacogenetics and Pharmacogenomics Studies

Currently, the most common therapeutic options for treating asthma include inhaled β2-adrenergic receptor agonists (beta agonists), inhaled corticosteroids (ICS), leukotriene modifiers, inhaled anticholinergics, and theophylline. In the near future, biologic therapies will also have an important place in the therapy of asthma. These drugs are classified into two groups according to the prevention of asthma or the symptom relief. The former groups are anti-inflammatory, especially inhaled corticosteroids and long-acting beta agonists (LABA). The second group includes drugs that act quickly relieving acute bronchoconstriction and symptoms like coughing, chest tightness, and wheezing, as short-acting beta agonists (SABA).

The ideal asthma treatment should be a tailored approach to health care in which adverse effects will be minimized and the therapeutic benefit for the patient is maximized. The clinical application of knowledge of patient's genetic profile can play a main role in this aspect.

3.1 β-Adrenergic Receptor Agonists

Inhaled beta agonists are the most commonly prescribed medical therapies for the management of asthma. Depending on the duration of their action, three types of inhaled beta agonists exist: the short-acting beta agonists (SABA: fenoterol, isoproterenol, pirbuterol, levalbuterol, and albuterol), the long-acting beta agonists (LABA: salmeterol and formoterol), and the ultra-long-acting beta agonists (vilanterol and indacaterol). The LABA therapy is administered in combination with an ICS as regular controller therapy, while SABA therapy is used for rescue, as-needed treatment for acute symptom relief or the prevention of exercise-induced symptoms. Despite the common use of these agents, this drug class is the center of a controversy related to concerns over adverse events. High doses of SABAs with less selective beta-adrenergic receptor activity were associated with serious adverse effects (including death) [17]. These drugs exert their action by activation of β2-adrenoreceptors (*ADRB2*) abundantly expressed in bronchial smooth muscle cells. The resulting of this activation leads to muscle relaxation, airway dilatation, and improved airflow [18]. In Table 1, genetic variants associated with response to β-adrenergic receptor agonist are shown.

3.1.1 Candidate Gene Analyses and Pathway Studies

ADRB2

The β2-adrenergic receptor (beta2-AR) is a member of the G-protein-coupled adrenergic receptor family with seven transmembrane segments; it was cloned by Kobilka et al. in 1987 and is localized to chromosome 5q31–q32, a region that has been linked with asthma and asthma-related phenotypes [19]. The *ADRB2* gene has been resequenced in multiple populations and more than 80 SNPs have been identified [20]. Two of these SNPs, Arg16Gly (rs1042713) and Glu27Gln (rs1042714), are the most common and have been well characterized in asthma pharmacogenetics [21]. The estimated frequency of the Arg16 variant is 39.3% in white Americans, 49.2% in blacks Americans, and 51.0% among Chinese [22].

With respect to correlation between this SNP and SABA, several studies demonstrated that Arg16 homozygotes had a greater acute response to SABA bronchodilation compared with Gly16 homozygotes, which has been confirmed in additional asthma populations [23, 24]. Several studies demonstrated higher FEV1

Table 1
Genetic variants associated with response to β-adrenergic receptor agonist

Gene	RsID	Study design	Drug class	Clinical association
ADRB2	rs1042713	Candidate gene	SABA	FEV1 bronchodilation PEFR response PEFR response
			LABA	PEFR response Bronchoprotection
	rs1800888	Candidate gene	LABA	Exacerbation asthma
ADCY9	rs2230739	Candidate gene	SABA LABA	FEV1 response
GSNOR	Promoter region	Candidate gene	Albuterol	Broncodilatador response
ARG1	rs2781659 rs2781667	Candidate gene	SABA	FEV1 bronchodilation
ARG2	rs7140310 rs10483801	Candidate gene	SABA	FEV1 bronchodilation
NOS3	rs1799983	Candidate gene	SABA	FEV1 bronchodilation
SPATS2L	rs295137	GWAS	SABA	FEV1 bronchodilation
SLC22A15	rs1281748 rs1281743	GWAS	SABA	FEV1 bronchodilation
SLC24A4	rs77441273	GWAS	SABA	FEV1 bronchodilation

(forced expiratory volume in the first second) increases after SABA (salbutamol) administration in homozygous Arg16 individuals as compared to heterozygous and homozygous Gly16 patients with polynosis [25, 26]. Patients who were homozygous Arg16 had impaired SABA response compared to homozygous Gly16 individuals [27]. Systematic administration of SABA to Arg16 asthmatics caused deterioration of lung function that did not stop even with treatment discontinuation. In contrast, patients homozygous for Gly16 demonstrated improved lung function [28, 29].

Some patients treated with salmeterol, experienced rare but severe asthma exacerbations [30]. Further investigation suggested dependence between Arg16 genotype and faster decline of lung parameters (FEV1) after LABA application [31]. However, certain controversy regarding adverse events has been reported [32, 33].

Regarding the relationship of this genetic variation in response to treatment in the pediatric population, it has been reported that Arg16 genotype increased susceptibility to exacerbations in children with asthma on regular LABA therapy. Montelukast may be proposed as tailored second-line controller therapy instead of salmeterol in asthmatic children [34]. The Pharmacogenetics of Asthma Medication in Children: Medication with Anti-inflammatory Effects (PACMAN) cohort study concluded that Arg16 homozygotes have a poorer response when treated with LABA and ICS [35]. This SNP at position 16 appears not to have clinical relevance to justify genotyping individuals [10].

In addition to these common SNPs, other less common, nonsynonymous coding variants have also been reported in the *ADRB2* gene. For example, the SNP rs1800888 encodes a Threonine to Isoleucine substitution at amino acid position 164 (Thr164Ile). The Ile164 isoform is three-to-four times less responsive to agonist-induced stimulation than carriers of the wild-type Thr164 [36]. Ortega et al. reported that Ile164 and -376ins were associated with adverse events during LABA therapy [37].

ADCY9

Adenylyl cyclase (*ADCY9*) belongs to the G-protein coupled β2-adrenergic receptor pathway. Polymorphisms in the gene encoding of this enzyme have been analyzed in several studies. Tantsira et al. [38] performed a study, in that SNP Ile772Met (rs2230739 in *ADCY9*) was associated with acute bronchodilatation in response to SABA in ICS-treated asthmatics from the CAMP study, while Kim et al. found that this SNP was also associated with lung function response to a LABA and ICS treatment in a Korean asthma population [39]. On the other hand, new studies conducted by Drake et al. have also shown that rare variants adjacent to *ADCY9* and *CRHR2* are associated with albuterol bronchodilator response in Puerto Rican and Mexican asthma subjects from the GALA (Genetics in Latino Americans) cohort [40].

GSNOR

Nitric oxide bioactivity, mediated through the formation of *S*-nitrosothiols (SNOs), has a significant effect on bronchomotor tone. *S*-nitrosoglutathione reductase (GSNOR) metabolizes *S*-nitrosothiol (SNO), an endogenous bronchodilator that is decreased in children with asthmatic respiratory failure and in adults with asthma undergoing segmental airway challenge [41]. Choudhry et al. recently sequenced the GSNOR gene in a subset asthmatics cohort and identified 13 SNPs with an allele frequency greater than 5% [42]. Association studies identified gene-gene interactions between *GSNOR* and *ADRB2* in Mexicans and Puerto Ricans combined, although these SNPs and their haplotypes were not associated with bronchodilator response.

ARG1 and *ARG2*

The arginases (ARG1 and ARG2) metabolize L-arginine, a natural substrate for nitric oxide synthase (NOS), to generate nitric oxide, which is an endogenous bronchodilator. Several recent studies support the role of *ARG1* and *ARG2* polymorphisms in bronchodilator response (BDR). Litonjua et al. reported that *ARG1* polymorphisms (rs2781659 and rs2781667) were associated with an acute SABA bronchodilator response in asthmatics [43]. These results have been replicated in independent asthma trial cohorts [44]. Vonk et al. conducted a study in a Dutch asthma cohort in that two polymorphisms in *ARG2* (rs17249437 and rs3742879) were associated with asthma and with more severe airway obstruction [45].

NOS3

Nitric oxide synthase enzymes have an important role in airway inflammation in asthmatic children. Iordanidou et al. investigated the association between NOS gene polymorphisms and response to ICS and LABA combination therapy demonstrating that the polymorphism Asp298Glu (rs1799983 in *NOS3*) was associated with lung function response to LABA and ICS [46].

3.1.2 GWASs Studies

SPATS2L

Himes et al. identified an SNP rs295137 close to the *SPATS2L* gene. Subsequently it has been suggested that SPATS2L may be a relevant regulator of β2-adrenergic receptor down regulation [47]. Another GWAS of BDR identified 4 SNPs (rs350729, rs1840321, rs1384918, and rs1319797) associated with BDR in chromosome 2 [48]. An intergenic SNP (rs11252394) was identified in the Childhood Asthma Management Program (CAMP) cohort-mediated GWAS of acute BDR to inhaled albuterol. This SNP is located proximal to several excellent biological candidates including protein kinase C (*PRKCQ*), interleukin receptors (*IL15RA*, *IL2RA*) among others [49].

SPATA13-AS1

Padhukasahasram et al. significantly associated *SPATA13-AS1* with SABA-induced bronchodilatation in healthy African Americans and replicated in two African American cohorts and one European American cohort with asthma [50].

SLC22A15

Drake KA et al. performed a GWAS with the objective of identifying genetic variation associated with BDR. Seven genetic variants associated with BDR were identified with significant threshold. Two rare variants were also identified in *SLC22A15* gene, being associated with increased BDR in Mexicans [40].

3.2 Inhaled Corticosteroids (ICS)

As previously commented, there are a group of patients less responsive to ICS, in which pharmacogenetic approaches might lead to improved personalized medicine [51]. The combination of wide interindividual variability and high intraindividual repeatability supports a genetic difference for the response to ICS in asthma [52]. In Table 2, genetic variants associated with response to inhaled corticosteroids are shown.

3.2.1 Candidate Gene Analyses and Pathway Studies

NR3C1

One of the earliest pharmacogenetic studies investigating glucocorticoids response involved the glucocorticoids receptor gene (*NR3C1*) located in chromosome 5q31. Two SNPs have demonstrated potential effects on corticosteroid response. The first is a relatively rare coding mutation, Asn363Ser (rs56149945 in *NR3C1*), which has been identified in several populations. Huizenga et al. conducted a study in which it was observed that lymphocytes from individuals with receptors with this genetic variation were shown to have a greater sensitivity to dexamethasone compared to lymphocytes from noncarriers [53].

Table 2
Genetic variants associated with response to inhaled corticosteroids

Gene	rs ID	Study design	Clinical association
NR3C1	rs56149945	Candidate gene	Sensitivity to drug
CRHR1	rs242941 rs1876828 rs242939	Candidate gene	Lung function FEV1 response
TBX21	rs2240017	Candidate gene	Bronchoprotection
STIP1	rs6591838 rs1011219	Candidate gene	FEV1 response
GLCCI1	rs37972	GWAS	FEV1 response
T gene	rs3127412 rs6456042	GWAS	FEV1 response
FBXL7	rs10044254	GWAS	ICS response
CA10	rs967676	GWAS	ICS response
SGK493	rs1440095	GWAS	ICS response
CTNNA3	rs1786929	GWAS	ICS response

The other genetic variation has potential pharmacogenetic implications. Gene expression studies have shown that the presence of the G allele in this mRNA stability motif increases the stability of GRβ mRNA, thus allowing increased rates of GRβ translation. Asthmatics possessing this G allele could therefore produce more GRβ, thus competing with functional GRα in the steroid dimer complex. These individuals could also have an attenuated response to exogenous and endogenous corticosteroids [5].

CRHR1

The *CRHR1* gene is the major receptor for corticotrophin, and it is the key regulator of corticosteroids synthesis and catecholamine production. The results of pharmacogenetic studies in this gene have been very promising. The most important pharmacogenetic studies have been conducted by Tantisira et al. [54]. This study found a significant correlation between lung function improvement after inhaled corticosteroid therapy and SNPs (rs1876828, rs242939, and rs242941). Homozygous individuals with this polymorphism had average FEV1 improvement, higher than homozygous patients lacking this SNP.

TBX21

Another candidate gene involved in ICS response is *TBX21*. *TBX21* knockout mice develop bronchial hyperresponsiveness, enhanced airway eosinophilia, and faster airway remodeling proving. The SNP His33Glu (rs2240017) has been associated with improvements in bronchial hyperresponsiveness or bronchoprotection. Tantisira et al. [55] conducted a study in the CAMP cohort that showed a significant decrease in airway hyperresponsiveness in heterozygous subjects for this SNP during ICS treatment as compared to His33His homozygous subjects and individuals that are not ICS-treated. This study was replicated in an independent Korean cohort with similar results [56].

FCER2

The *FCER2* gene encodes the low-affinity receptor for IgE, a key molecule for B-cell activation and growth. A novel variant in *FCER2* has recently been associated with asthma exacerbations while on ICS The SNP, rs28364072, was associated with increased risk of exacerbations in asthmatic children taking ICS [57]. This novel variant was also associated with both higher IgE levels and with differential expression of the *FCER2* gene, supporting the contention that variation in *FCER2* can adversely affect normal negative feedback in the control of IgE synthesis and action. Others studies have been also confirmed these results [58, 59].

STIP1

Multiple genes encode for the heterocomplex of chaperones and immunophilins that bind the glucocorticoids receptor and mediate proper assembly and activation of the receptor. Genetic variations in the gene encoding for the heat shock organizing protein (STIP1) have been associated with the regulation of corticosteroid response in asthmatic subjects with reduced lung function. Thus, Hawkins et al. found that the SNPs (rs6591838, rs2236647, and rs1011219 in *STIP1*) were significantly associated with improvement in FEV1 response after 4 or 8 weeks of corticosteroid therapy [60].

3.2.2 GWASs Studies

Investigations into the genetic basis for treatment response heterogeneity in asthma have largely focused on candidate genes. These candidate gene pharmacogenetic investigations have identified associations related to the variable response to corticosteroids as measured by differences in lung function, airways responsiveness, bronchodilator response, and exacerbations. However, progress related to identifying sufficient numbers of variants to achieve the predictive medicine for any ICS treatment phenotype through the use of candidate genes has been slow. GWAS provides the ability to rapidly identify novel pharmacogenetic variants by simultaneously interrogating genetic variants from across the genome. The following describe the most important GWASs studies to date:

GLCCI1

A recent study demonstrated an SNP in the promoter region of the glucocorticoids-induced transcript 1 gene (*GLCCI1*), rs37972, which was associated with lung function responses to inhaled glucocorticoids. The SNP rs37972 is also in strong linkage equilibrium (i.e., strongly correlated or "tagged") with another *GLCI1* promoter SNP, rs37973, which determines gene transcription in vitro, demonstrating a functional or molecular-based rationale for the observed genetic effects of variation in this gene on corticosteroid response. It will be important to replicate these corticosteroid response gene variants in other, larger populations and determine whether they are independent predictors or have additive effects that regulate corticosteroid responses in asthma [61].

T Gene

Another GWAS using the SHARP cohort identified variants in the *T* gene (T, brachyury homolog) (rs3127412, rs6456042) [62].

FBXL7

Park et al. performed the first GWAS of ICS response. Three SNPs were found but were not replicated in adult cohorts; rs10044254 was found in intronic region of *FBXL7* gene and is associated with decreased expression in immortalized B cell derived from CAMP participants. Author suggested that specific mechanism regulating symptomatic response to ICSs in children does not work in adults [63].

CA10, SGK493, and CTNNA3

Perin and Potĉnick have performed a GWAS in different phenotypes of childhood asthma. Authors reported significantly associations between *CA10* (rs967676), *SGK493* (rs1440095), and *CTNNA3* (rs1786929) with asthma and glucocorticoids response [64].

Wang et al. identified rs6924808, rs10481450, rs1353649, rs12438740, and rs2230155 significantly associated with corticosteroid dose–response. This report is the first and unique with this pharmacodynamic approach, and has been revealed to be statistically more powerful for gene detection compared with classical single-dose approaches [65].

Himes et al. performed an RNAseq analysis to characterize transcriptomic changes in four primary human airway smooth muscle (ASM) cells in response to dexamethasone (potent synthetic glucocorticoids). Authors identified *CRISPLD2*, which encodes a secreted protein previously implicated in lung development and endotoxin regulation. It was found to have SNPs associated with inhaled corticosteroid resistance and bronchodilator response among asthma patients. Quantitative RT-PCR and Western blotting showed that dexamethasone treatment significantly increased *CRISPLD2* mRNA and protein expression in ASM cells. Functional studies revealed that *CRISPLD2* is an asthma pharmacogenetics candidate gene that regulates anti-inflammatory effects of glucocorticoids in the ASM [15].

Another integrative approach combines an eQTL analysis with GWAS data and expression microarrays from lymphoblastoid cell lines in corticosteroid treated and untreated cells. Multiple novel genome-wide significant pharmacogenomic loci were identified in both Caucasian and African Americans including rs6504666 and rs1380657 (*SPATA20*), rs12891009 (*ACOT4*), rs2037925 and rs2836987 (*BRWD1*), rs1144764 (*ALG8*), and rs3793371 (*NAPRT1*) [66].

Table 3
Genetic variants associated with response to Montelukast

Gene	rs ID	Study design	Clinical associations
ALOX5	Promoter repeat rs892690 rs2029253 rs2115819	Candidate gene	Changes FEV1 and PEF Exacerbation risk
LTC4S	rs730012	Candidate-gene	Exacerbation risk
CYSLTR2	Rs912278	Candidate gene	Changes in PEF
ABCC1	rs119774	Candidate gene	Changes in FEV1
LTA4H	rs2660845	Candidate gene	Decrease of FEV1 and PEF
SLCO2B1	rs12422149	Candidate gene	Decrease of serum drug levels
MRPP3	rs12436663	GWAS	Decrease of FEV1
GLT1D1	rs517020	GWAS	Poor response
MLLT3	rs6475448	GWAS	Improve response

3.3 Leukotriene Modifiers

Leukotriene modifiers play an important role in a variety of inflammatory diseases such as asthma [58]. This drug class shows strong anti-inflammatory activity, ameliorate asthma clinical course and improve disease control with minimal or no side effects. Currently, based on their mechanism of action, there are two groups of drugs: cysteine leukotrienes receptor antagonists (montelukast, zafirlukast, pranlukast, and tomelukast) and 5-lipoxygenase inhibitors (zileuton). In Table 3, genetic variants associated with response to leukotriene modifiers are shown.

3.3.1 Genes Candidates Analyses and Pathway Studies

To date, most pharmacogenetics investigations, which may affect therapy with leukotriene modifiers, have focused on variants of *ALOX5* and *LTC4S*. Other genetic variants of *LTA4H*, *MRP1*, and *CYSLTR1* have also been investigated.

ALOX5

The *5-LOX* gene (*ALOX5*) is located on chromosome 10q11.12, contains 14 exons, and its activity is associated with a number of repetitions of Sp1/Erg1 binding motifs in the promoter region. Mutant *ALOX5* repeat polymorphism was associated with decreased exacerbation rates in white asthmatic patients treated with montelukast [67], while other studies show a decrease number of asthma exacerbations, improvement of forced expiratory volume at 1 s (FEV1), and decreased use of beta2 agonists in Spanish patients with wild-type allele or heterozygous variant allele [68] or no significant difference in terms of bronchodilator response or bronchial hyperresponsiveness between wild-type allele and heterozygous variant allele [69].

In a recent genetic study of the *ALOX5* promoter, variant in homozygotes for the less common repeat alleles had a higher urinary concentration of leukotriene E4, consistent with increased leukotriene biosynthesis. Most importantly, 86% of these minor risk alleles were from African American subjects suggesting that genetic variation from an African ancestry may influence asthma severity and responsiveness to therapies targeting the leukotriene pathway [70]. Subsequent larger candidate gene studies have shown that additional *ALOX5* SNPs (rs2115819, rs4987105, and rs4986832) may also influence the response to the leukotriene receptor antagonist, montelukast [67, 71].

LTC4S

LTC4 synthase (LTC4S) belongs to S-glutathione synthases family and catalyzes the conversion of LTA4 to LTC4. The most important polymorphism detected is the A-444C (rs730012), which is implicated in an increase in the production of LTC4 in eosinophils. Sanak et al. confirm that this genetic variant occurs more often in patients with aspirin idiosyncrasy [72]. While Lima et al. found that in heterozygous individuals for this variation had a 73% lower risk of asthma exacerbations compared to homozygous individuals during the treatment with montelukast [67]. Other studies performed in Southern Europe [73] and Japan [74] also reported similar results.

CYSLTR1 and CYSLTR2

The CYSLT1 and CYSLT2 receptors have been characterized as G-protein-coupled receptors. Some studies have examined the relationship of polymorphisms in these genes with leukotriene modifiers, but these results are not very promising. Tansira et al. conducted a study on patients treated with zileuton, but not significant correlation could be established between *CYSLT1* gene polymorphisms and clinical response to therapy [75]. However, Klotsmant et al. performed a study in which a strongest statistical evidence of clinically relevant pharmacogenetic effects peak expiratory flow was identified in *CYSLTR2* (rs91227 and rs912278) [71].

ABCC1

The *ABCC1* gene encodes MRP1 (Multiple Drug Resistance Protein 1), which plays an important role in transmembrane transport of LTC4. One of the polymorphisms of this gene (rs119774 in *LTC4*) had been correlated to montelukast response. Lima et al. observed that in heterozygous individuals for this polymorphism had a 24% FEV1 rise as compared to only a 2% improvement in homozygous individuals [67].

LTA4

LTA4 hydrolase is an enzyme that converts LTA4 to LTB4. The gene encoding this protein is located on chromosome 12q22. One of the known polymorphisms for this gene (rs2660845 in *LTA4*) has been associated with risk of asthma exacerbation.

Heterozygous individuals for this SNP have 4–5 higher risk of asthma exacerbation than homozygous individuals, during treatment with montelukast. The pathogenic mechanism of this process remains unclear. It has been hypothesized that this SNP causes a decreased enzyme activity that results in diminished LTB4 synthesis, therefore stimulating the LTC4-synthase pathway and leading to cysteine leukotriene synthesis [67].

SLCO2B1

The gene *SLCO2B1* encodes the protein 2B1 that plays an important role in the active transport of organic anions through the intestinal wall. SLCO2B1 gene polymorphisms have been studied in order to establish their relationship with montelukast pharmacokinetics. The most important polymorphism described is Arg312Gln (rs12422149), which is involved in the transport and serum levels of montelukast. Mougey et al. demonstrated that individuals with this SNP have a significantly lower serum drug concentration [76]. However, these effects of polymorphisms of the *SLCO2B1* transporter gene on the pharmacokinetics of montelukast have not been reproduced in smaller studies [77, 78].

3.3.2 GWASs Studies

The first GWAS of the leukotriene modifier response in asthma has been recently published. The study analyzed DNA and phenotypic information from two placebo-controlled trials of zileuton response evaluating change in FEV1 following leukotriene modifiers treatment. The top 50 SNP associations were replicated in an independent zileuton cohort, and two cohorts of montelukast response. The rs12436663 in *MRPP3* was significantly associated. The homozygous carriers showed a significant reduction in FEV1 change following zileuton treatment. The rs517020 in *GLT1D1* was associated with poor response after montelukast and zileuton treatment [79]. A second GWAS was performed using phenotypic data available from American Lung Association-Asthma Clinical Research Center (ALA-ACRC) cohorts. A novel pharmacogenomic locus (rs6475448) in *MLLT3* was identified associated to improved montelukast response in asthmatics [80].

4 Future Perspectives

In last years, human genetic research has made a considerable progress in both genome analysis techniques and international research networks for human genome information (consortiums). Pharmacogenetics has evolved into pharmacogenomics by the application of GWASs, next-generation sequencing, and newer integrative techniques to study drug response phenotypes. Large-scale whole-exome and whole-genome sequencing projects such as the National Institutes of Health National Heart, Lung, and Blood Institute GO Exome Sequencing Program, the 1000 Genomes

Project and the Consortium of Asthma in African Ancestry Populations have resulted in an excellent database of common and rare genetic variants, for future pharmacogenetic studies in different racial and ethnic groups [81, 82].

Many current studies are limited by their small sample size, imprecise phenotype definitions, population stratification, and lack of replication of the results. Future studies should focus on these limitations. However, other factors, such as gene-environment interactions, epigenetic regulation, and interactions between variants in different genes and genetic pathways, DNA methylation, Histone modifications, transcriptional regulation by small interfering RNA (siRNA), micro RNA (miRNA), and long non-coding RNA (lncRNA), are also being explored as biomarkers in therapeutics interventions [83–85]. In addition to molecular genetics studies, functional studies in cellular models or animals as shown in this book are also necessary for studying variation in the expression of the different polymorphic genes. Another strategy for drug development is a biological system approach, focusing in multiple omics like metabolomics and pharmaco-proteomics studies of an animal model [10]. Translating the findings into clinical practice is the main challenge to scientific. Some of the identified genetic variations havelarge effects and predict drug responses. These pharmacogenomics biomarkers may determine which patients will benefit from a specific drug. The cost of high-throughput technologies are decreasing very fast and it is expected that in near future, these should be proposed for use in clinical trials and this user-friendly technology should help to test individual patients.

References

1. Moore WC, Meyers DA, Wenzel SE et al (2010) Identification of asthma phenotypes using cluster analysis in the Severe Asthma Research Program. Am J Respir Crit Care Med 181(4):315–323
2. Miranda C, Busacker A, Balzar S et al (2004) Distinguishing severe asthma phenotypes: role of age at onset and eosinophilic inflammation. J Allergy Clin Immunol 113(1):101–108
3. Fanta CH (2009) Asthma. N Engl J Med 360(10):1002–1014
4. Masoli M, Fabian D, Holt S et al (2004) The global burden of asthma: executive summary of the GINA Dissemination Committee report. Allergy 59(5):469–478
5. Hawkins GA, Peters SP (2008) Pharmacogenetics of asthma. Methods Mol Biol 448:359–378
6. García-Sánchez A, Isidoro-García M, García-Solaesa V et al (2015) Genome-wide association studies (GWAS) and their importance in asthma. Allergol Immunopathol (Madr) 43:601–608. doi:10.1016/j.aller.2014.07.004
7. Lorente F, Isidoro-Garcia M, Macias E et al (2010) Do genetic factors determine atopy or allergy? Allergol Immunopathol (Madr) 38(2):53–55
8. Vogel FVF (1959) Modern problems of human genetics. Ergeb Inn Med Kinderheilkd 12:52–125
9. Wang L, McLeod HL, Weinshilboum RM (2011) Genomics and drug response. N Engl J Med 364(12):1144–1153
10. Davis JS, Weiss ST, Tantisira KG (2015) Asthma pharmacogenomics: 2015 update. Curr Allergy Asthma Rep 15(7):42
11. Patnala R, Clements J, Batra J (2013) Candidate gene association studies: a comprehensive guide to useful in silico tools. BMC Genet 14:39

12. Kwon JM, Goate AM (2000) The candidate gene approach. Alcohol Res Health 24(3): 164–168
13. Karczewski KJ, Daneshjou R, Altman RB (2012) Chapter 7: pharmacogenomics. PLoS Comput Biol 8(12):e1002817
14. Ober C, Yao TC (2011) The genetics of asthma and allergic disease: a 21st century perspective. Immunol Rev 242(1):10–30
15. Himes BE, Jiang X, Wagner P et al (2014) RNA-Seq transcriptome profiling identifies CRISPLD2 as a glucocorticoid responsive gene that modulates cytokine function in airway smooth muscle cells. PLoS One 9(6):e99625
16. Michaelson JJ, Loguercio S, Beyer A (2009) Detection and interpretation of expression quantitative trait loci (eQTL). Methods 48(3): 265–276
17. Ortega VE, Meyers DA, Bleecker ER (2015) Asthma pharmacogenetics and the development of genetic profiles for personalized medicine. Pharmgenomics Pers Med 8:9–22
18. Litonjua AA, Gong L, Duan QL et al (2010) Very important pharmacogene summary ADRB2. Pharmacogenet Genomics 20(1): 64–69
19. Kobilka BK, Dixon RA, Frielle T et al (1987) cDNA for the human beta 2-adrenergic receptor: a protein with multiple membrane-spanning domains and encoded by a gene whose chromosomal location is shared with that of the receptor for platelet-derived growth factor. Proc Natl Acad Sci U S A 84(1):46–50
20. Weiss ST, Litonjua AA, Lange C et al (2006) Overview of the pharmacogenetics of asthma treatment. Pharmacogenomics J 6(5):311–326
21. Reihsaus E, Innis M, MacIntyre N et al (1993) Mutations in the gene encoding for the beta 2-adrenergic receptor in normal and asthmatic subjects. Am J Respir Cell Mol Biol 8(3): 334–339
22. Maxwell TJ, Ameyaw MM, Pritchard S et al (2005) Beta-2 adrenergic receptor genotypes and haplotypes in different ethnic groups. Int J Mol Med 16(4):573–580
23. Lima JJ, Thomason DB, Mohamed MH et al (1999) Impact of genetic polymorphisms of the beta2-adrenergic receptor on albuterol bronchodilator pharmacodynamics. Clin Pharmacol Ther 65(5):519–525
24. Choudhry S, Ung N, Avila PC et al (2005) Pharmacogenetic differences in response to albuterol between Puerto Ricans and Mexicans with asthma. Am J Respir Crit Care Med 171(6):563–570
25. Woszczek G, Borowiec M, Ptasinska A et al (2005) Beta2-ADR haplotypes/polymorphisms associate with bronchodilator response and total IgE in grass allergy. Allergy 60(11): 1412–1417
26. Martinez FD, Graves PE, Baldini M et al (1997) Association between genetic polymorphisms of the beta2-adrenoceptor and response to albuterol in children with and without a history of wheezing. J Clin Invest 100(12): 3184–3188
27. Carroll CL, Stoltz P, Schramm CM et al (2009) Beta2-adrenergic receptor polymorphisms affect response to treatment in children with severe asthma exacerbations. Chest 135(5):1186–1192
28. Israel E, Drazen JM, Liggett SB et al (2000) The effect of polymorphisms of the beta(2)-adrenergic receptor on the response to regular use of albuterol in asthma. Am J Respir Crit Care Med 162(1):75–80
29. Israel E, Chinchilli VM, Ford JG et al (2004) Use of regularly scheduled albuterol treatment in asthma: genotype-stratified, randomised, placebo-controlled cross-over trial. Lancet 364(9444):1505–1512
30. Nelson HS, Weiss ST, Bleecker ER et al (2006) The Salmeterol Multicenter Asthma Research Trial: a comparison of usual pharmacotherapy for asthma or usual pharmacotherapy plus salmeterol. Chest 129(1):15–26
31. Wechsler ME, Lehman E, Lazarus SC et al (2006) Beta-Adrenergic receptor polymorphisms and response to salmeterol. Am J Respir Crit Care Med 173(5):519–526
32. Wechsler ME, Kunselman SJ, Chinchilli VM et al (2009) Effect of beta2-adrenergic receptor polymorphism on response to long acting beta2 agonist in asthma (LARGE trial): a genotype-stratified, randomized, placebo-controlled, crossover trial. Lancet 374(9703):1754–1764
33. Bleecker ER et al (2010) Beta2-receptor polymorphisms in patients receiving salmeterol with or without fluticasone propionate. Am J Respir Crit Care Med 181(7):676–687
34. Lipworth BJ, Basu K, Donald HP et al (2013) Tailored second-line therapy in asthmatic children with the Arg(16) genotype. Clin Sci (Lond) 124(8):521–528
35. Zuurhout MJ, Vijverberg SJ, Raaijmakers JA et al (2013) Arg16 ADRB2 genotype increases the risk of asthma exacerbation in children with a reported use of long-acting beta2-agonists: results of the PACMAN cohort. Pharmacogenomics 14(16):1965–1971
36. Green SA, Cole G, Jacinto M et al (1993) A polymorphism of the human beta 2-adrenergic receptor within the fourth transmembrane domain alters ligand binding and functional

properties of the receptor. J Biol Chem 268(31):23116–23121

37. Ortega VE, Hawkins GA, Moore WC et al (2014) Effect of rare variants in ADRB2 on risk of severe exacerbations and symptom control during long acting beta agonist treatment in a multiethnic asthma population: a genetic study. Lancet Respir Med 2(3):204–213
38. Tantisira KG, Small KM, Litonjua AA et al (2005) Molecular properties and pharmacogenetics of a polymorphism of adenylyl cyclase type 9 in asthma: interaction between beta-agonist and corticosteroid pathways. Hum Mol Genet 14(12):1671–1677
39. Kim SH, Ye YM, Lee HY et al (2011) Combined pharmacogenetic effect of ADCY9 and ADRB2 gene polymorphisms on the bronchodilator response to inhaled combination therapy. J Clin Pharm Ther 36(3):399–405
40. Drake KA, Torgerson DG, Gignoux CR et al (2014) A genome-wide association study of bronchodilator response in Latinos implicates rare variants. J Allergy Clin Immunol 133(2):370–378
41. Que LG, Yang Z, Stamler JS et al (2009) S-nitrosoglutathione reductase: an important regulator in human asthma. Am J Respir Crit Care Med 180(3):226–231
42. Choudhry S, Que LG, Yang Z et al (2010) GSNO reductase and beta2-adrenergic receptor gene-gene interaction: bronchodilator responsiveness to albuterol. Pharmacogenet Genomics 20(6):351–358
43. Litonjua AA, Lasky-Su J, Scheiter K et al (2008) ARG1 is a novel bronchodilator response gene: screening and replication in four asthma cohorts. Am J Respir Crit Care Med 178(7):688–694
44. Duan QL, Gaume BR, Hawkins GA et al (2011) Regulatory haplotypes in ARG1 are associated with altered bronchodilator response. Am J Respir Crit Care Med 183(4):449–454
45. Vonk JM, Postma DS, Maarsingh H et al (2010) Arginase 1 and arginase 2 variations associate with asthma, asthma severity and beta2 agonist and steroid response. Pharmacogenet Genomics 20(3):179–186
46. Iordanidou M, Paraskakis E, Tavridu A et al (2012) G894T polymorphism of eNOS gene is a predictor of response to combination of inhaled corticosteroids with long-lasting beta2-agonists in asthmatic children. Pharmacogenomics 13(12):1363–1372
47. Himes BE, Jiang X, Hu R et al (2012) Genome-wide association analysis in asthma subjects identifies SPATS2L as a novel bronchodilator response gene. PLoS Genet 8(7): e1002824
48. Israel E, Lasky-Su J, Markezich A et al (2015) Genome-wide association study of short-acting beta2-agonists. A novel genome-wide significant locus on chromosome 2 near ASB3. Am J Respir Crit Care Med 191(5):530–537
49. Duan QL, Laskky-Su J, Himes BE et al (2014) A genome-wide association study of bronchodilator response in asthmatics. Pharmacogenomics J 14(1):41–47
50. Padhukasahasram BK, Yang JJ, Levin AM et al (2014) Gene-based association identifies SPATA13-AS1 as a pharmacogenomic predictor of inhaled short-acting beta-agonist response in multiple population groups. Pharmacogenomics J 14(4):365–371
51. Ortega VE, Bleecker ER (2012) The pharmacogenetics of asthma and the road to personalized medicine. Pulmão RJ 21(2):41–52
52. Lima JJ, Blake KV, Tantisira KG et al (2009) Pharmacogenetics of asthma. Curr Opin Pulm Med 15(1):57–62
53. Huizenga NA, Koper JW, De Lange P et al (1998) A polymorphism in the glucocorticoid receptor gene may be associated with and increased sensitivity to glucocorticoids in vivo. J Clin Endocrinol Metab 83(1):144–151
54. Tantisira KG, Lake S, Silverman ES (2004) Corticosteroid pharmacogenetics: association of sequence variants in CRHR1 with improved lung function in asthmatics treated with inhaled corticosteroids. Hum Mol Genet 13(13): 1353–1359
55. Tantisira KG, Hwang ES, Raby BA (2004) TBX21: a functional variant predicts improvement in asthma with the use of inhaled corticosteroids. Proc Natl Acad Sci U S A 101(52): 18099–18104
56. Ye YM, Lee HY, Kim SH et al (2009) Pharmacogenetic study of the effects of NK2R G231E G>A and TBX21 H33Q C>G polymorphisms on asthma control with inhaled corticosteroid treatment. J Clin Pharm Ther 34(6):693–701
57. Tantisira KG, Silverman ES, Mariani TJ et al (2007) FCER2: a pharmacogenetic basis for severe exacerbations in children with asthma. J Allergy Clin Immunol 120(6):1285–1291
58. Tse SM, Tantisira K, Weiss ST (2011) The pharmacogenetics and pharmacogenomics of asthma therapy. Pharmacogenomics J 11(6): 383–392
59. Maitland-van der Zee AH, Raaijmakers JA (2012) Variation at GLCCI1 and FCER2: one step closer to personalized asthma treatment. Pharmacogenomics 13(3):243–245
60. Hawkins GA, Lazarus R, Smith RS et al (2009) The glucocorticoid receptor heterocomplex

gene STIP1 is associated with improved lung function in asthmatic subjects treated with inhaled corticosteroids. J Allergy Clin Immunol 123(6):1376–1383.e7

61. Tantisira KG, Lasky-Su J, Harada M et al (2011) Genomewide association between GLCCI1 and response to glucocorticoid therapy in asthma. N Engl J Med 365(13):1173–1183
62. Tantisira KG, Damask A, Szefler SJ et al (2012) Genome-wide association identifies the T gene as a novel asthma pharmacogenetic locus. Am J Respir Crit Care Med 185(12):1286–1291
63. Park HW, Dahlin A, Tse S et al (2014) Genetic predictors associated with improvement of asthma symptoms in response to inhaled corticosteroids. J Allergy Clin Immunol 133(3): 664–669.e5
64. Perin P, Potocnik U (2014) Polymorphisms in recent GWA identified asthma genes CA10, SGK493, and CTNNA3 are associated with disease severity and treatment response in childhood asthma. Immunogenetics 66(3):143–151
65. Wang Y, Tong C, Wang Z et al (2015) Pharmacodynamic genome-wide association study identifies new responsive loci for glucocorticoid intervention in asthma. Pharmacogenomics J 15(5):422–429
66. Qiu W, Rogers AJ, Damask A et al (2014) Pharmacogenomics: novel loci identification via integrating gene differential analysis and eQTL analysis. Hum Mol Genet 23(18): 5017–5024
67. Lima JJ, Zhang S, Grant A et al (2006) Influence of leukotriene pathway polymorphisms on response to montelukast in asthma. Am J Respir Crit Care Med 173(4):379–385
68. Telleria JJ, Blanco-Quiros A, Varillas D et al (2008) ALOX5 promoter genotype and response to montelukast in moderate persistent asthma. Respir Med 102(6):857–861
69. Fowler SJ, Hall IP, Wilson AM et al (2002) 5-Lipoxygenase polymorphism and in-vivo response to leukotriene receptor antagonists. Eur J Clin Pharmacol 58(3):187–190
70. Mougey E, Lang JE, Allayee H (2013) ALOX5 polymorphism associates with increased leukotriene production and reduced lung function and asthma control in children with poorly controlled asthma. Clin Exp Allergy 43(5):512–520
71. Klotsman M, York TP, Pillai SG et al (2007) Pharmacogenetics of the 5-lipoxygenase biosynthetic pathway and variable clinical response to montelukast. Pharmacogenet Genomics 17(3):189–196
72. Sanak M, Simon HU, Szczeklik A (1997) Leukotriene C4 synthase promoter polymorphism and risk of aspirin-induced asthma. Lancet 350(9091):1599–1600
73. Isidoro-Garcia M, Davila I, Moreno E et al (2005) Analysis of the leukotriene C4 synthase A-444C promoter polymorphism in a Spanish population. J Allergy Clin Immunol 115(1): 206–207
74. Asano K, Shiomi T, Hasegawa N et al (2002) Leukotriene C4 synthase gene A(-444)C polymorphism and clinical response to a CYS-LT(1) antagonist, pranlukast, in Japanese patients with moderate asthma. Pharmacogenetics 12(7):565–570
75. Tantisira KG, Lima J, Sylvia J et al (2009) 5-lipoxygenase pharmacogenetics in asthma: overlap with Cys-leukotriene receptor antagonist loci. Pharmacogenet Genomics 19(3):244–247
76. Mougey EB, Feng H, Castro M et al (2009) Absorption of montelukast is transporter mediated: a common variant of OATP2B1 is associated with reduced plasma concentrations and poor response. Pharmacogenet Genomics 19(2):129–138
77. Tapaninen T, Karonen T, Backman JT et al (2013) SLCO2B1 c.935G>A single nucleotide polymorphism has no effect on the pharmacokinetics of montelukast and aliskiren. Pharmacogenet Genomics 23(1):19–24
78. Kim KA, Lee HM, Joo HJ et al (2013) Effects of polymorphisms of the SLCO2B1 transporter gene on the pharmacokinetics of montelukast in humans. J Clin Pharmacol 53(11):1186–1193
79. Dahlin A, Litonjua A, Irvin CG et al (2015) Genome-wide association study of leukotriene modifier response in asthma. Pharmacogenomics J. doi: 10.1038/tpj.2015.34
80. Dahlin A, Litonjua A, Lima JJ et al (2015) Genome-wide association study identifies novel pharmacogenomic loci for therapeutic response to montelukast in asthma. PLoS One 10(6): e0129385
81. Ortega VE, Meyers DA (2014) Pharmacogenetics: implications of race and ethnicity on defining genetic profiles for personalized medicine. J Allergy Clin Immunol 133(1):16–26
82. Abecasis GR, Auto A, Brooks LD et al (2012) An integrated map of genetic variation from 1,092 human genomes. Nature 491(7422):56–65
83. Sessa R, Hata A (2013) Role of microRNAs in lung development and pulmonary diseases. Pulm Circ 3(2):315–328
84. Perry MM, Tsitsiou E, Austin PJ et al (2014) Role of non-coding RNAs in maintaining primary airway smooth muscle cells. Respir Res 15:58
85. Booton R, Lindsay MA (2014) Emerging role of MicroRNAs and long noncoding RNAs in respiratory disease. Chest 146(1):193–204

Index

A

B

C

D

María Isidoro-García (ed.), *Molecular Genetics of Asthma*, Methods in Molecular Biology, vol. 1434,
DOI 10.1007/978-1-4939-3652-6, © Springer Science+Business Media New York 2016

O

P

Q

R

S

T

U

V

W